The New Biology

Law, Ethics, and Biotechnology

George P. Smith II, J.D.
The Catholic University of America
School of Law
Washington, D.C.

Plenum Press • New York and London

Library of Congress Cataloging in Publication Data

Smith, George Patrick, 1939–
    The new biology: law, ethics, and biotechnology / George P. Smith II.
        p.   cm.
    Includes bibliographies and index.
    ISBN 0-306-43187-4
    1. Biotechnology—Moral and ethical aspects. 2. Genetic engineering—Moral
and ethical aspects. 3. Medical ethics. I. Title.
    [DNLM: 1. Biotechnology. 2. Ethics, Medical. 3. Jurisprudence. W. 50 S648n]
TP248.2.S65   1989
174'.25—dc19
DNLM/DLC   88-9482
for Library of Congress   CIP

© 1989 Plenum Press, New York
A Division of Plenum Publishing Corporation
233 Spring Street, New York, N.Y. 10013

Printed in the United States of America

Dedicated to

My valued friend and colleague, Dr. John L. Garvey,
Whiteford Professor of Law and Dean Emeritus, The
Catholic University School of Law, *clarum est*
venerabile nomen, with respect, admiration, and
enduring appreciation.

Preface

Improvement of man's genetic endowment by direct actions aimed at striving for the positive propagation of those with a superior genetic profile (an element of which is commonly recognized as a high intelligence quotient) or—conversely—delimitation of those with negative genetic inheritance has always remained a primary concern of the geneticist and the social engineer. Genetic integrity, eugenic advancement, and a strong genetic pool designed to eliminate illness and suffering have been the benchmarks of the "Genetic Movement" and the challenge of Orwell's *Nineteen Eighty-Four*.

If the quality of life can in some way be either improved or advanced by use of the law, then this policy must be developed and pursued. No longer does the Dostoyevskian quest to give life meaning through suffering become an inescapable given. By and through the development and application of new scientific advances in the field of genetics (and especially genetic engineering), the real potential exists to prevent, to a very real extent, most human suffering before it ever manifests itself in or through life. Freedom to undertake research in the exciting and fertile frontiers of the "New Biology" and to master the Genetic Code must be nurtured and maintained. The search for the truth inevitably prevents intellectual, social, and economic stagnation, as well as—ideally—frees all from anxiety and fright. Yet, there is a very real potential for this quest to confuse and confound. It is for man himself to decide which he allows to dominate or direct the course of his future investigations and subsequent actions.

The basic challenge of the New Biology, then, is to seek and maintain quality in purposeful living, both in the early potential for life and its subsequent continuation, yet at the same time protect the recognition of sanctity of life—again, at its conception through its natural conclusion. What is often seen as a consequence of unique factual circumstance is a balancing of one goal against the other. The situation ethic is predominant over an a priori standard. Viewed from yet another perspective, this balancing test underscores recognition of the fact that human life is, in actuality, but a resource, just as are natural, physical, and environmental resources. Considered as such, a goal for the conservation of every resource is the maximization of its use or potential, be it economic, social, cultural, or political. Waste, or the nonproductive utilization of the resource or asset, must be avoided. Thus, in seeking to maximize the good of this previous resource of life, the right of personal autonomy or self-determination and spiritual awareness are vectors of force that must be factored additionally into any balancing equation. State interest is still another positive force and also a constraint on decision making here. The state exists to promote the public good or the public welfare as it perceives it to be.

Surely, one aspect of this promotion is the minimization of physical suffering.

In the "brave new world" of tomorrow—which is, in many respects, already here—biomedical and biotechnological decisions will of necessity be involved with volatile, sensitive, and highly charged areas of autonomy, procreation, and life, death, and immortality, as well as the freedom of scientific inquiry and the role of religion and ethics. What will become manifest through the chapters that follow is the realization that a test (sometimes active, at other times inactive) is effected in order to bring about a resolution of a particular controversy enmeshed with ethical conundrums. My thesis is that the test will seek to weigh the utility of the good (socioeconomic, cultural, or political) of maintaining the status quo against the gravity of the harm of undertaking a new and different course of action—be it efforts to obtain organs for human transplantation or to allocate scarce medical resources to maintain terminally ill individuals and those with a prolonged and expensive course of treatment, such as AIDS patients.

The simple yet sometimes elusive goal of any deliberative process involving the technologies of the present and its attendant difficulties as well as the future should always be to maximize the total potential for human growth, development, interpersonal relations, and intellectual fulfillment when it exists, and at the same time minimize all suffering connected with the attainment and perpetuation of this lifetime goal. In the final analysis, then, these chapters explore some of the truly great challenges that are in turn great opportunities for mastery of the New Biology; a mastery that will not only serve to advance scientific knowledge but combat disease and thereby increase the overall standard of qualitative living.

These chapters are extended versions, in several cases, of the 1984 Fulbright Lectures on Law and Medical Jurisprudence, entitled *1984: A Brave or a Confused New World?*, which I gave throughout the continent of Australia during the time I was the Fulbright Visiting Professor of Law and Medical Jurisprudence at the Faculty of Law, University of New South Wales, Australia. During the summer of 1987, I returned to Sydney, Australia, where I presented the Julius Stone Memorial Oration, *The Province and Function of Law, Science, and Medicine*, and also presented a set of lectures entitled *Opportunities of the New Biology—Perplexities in the Ying and the Yang*, at the National University of Singapore. Significant parts of these lectures are also included in this collection.

George P. Smith II

Washington, D.C.

Acknowledgments

I wish to acknowledge, with gratitude, permission to draw upon and reprint parts of the following copyrighted materials which I have authored previously:

Australia's Frozen Orphan Embryos: A Medical, Legal and Ethical Dilemma, 24 *Journal of Family Law* 27 (1985).

Sexuality, Privacy and the New Biology, 67 *Marquette Law Review* 63 (1984).

Quality of Life, Sanctity of Creation: Palliative or Apotheosis? 63 *Nebraska Law Review* 707 (1984).

The Promise of Abundant Life: Patenting a Magnificent Obsession, 8 *Journal of Contemporary Law* 85 (1982).

Contents

Chapter 6

*El Dorado and the Promise of Cryonic
Suspension*

Chapter 7

AIDS: The Private and the Public Dilemmas

Chapter 8

Noble Death, Rational Suicide, or
Self-Determination 153

Chapter 9

Procreational Autonomy: Values Gone Awry? 171

Chapter 10

The Case of the Orphan Embryos   199

Chapter 11

Science, Religion, and the New Biology   209

Chapter 1

Biotechnology

The Challenges and the Opportunities

Although 1984 has come and gone, the Orwellian spectres to which it gave rise grow almost daily into realities.[1] Futurology is a study, ideology, or movement that advances beliefs in a future era of unrestricted progress advanced through the harnessing of science and technology for the betterment or transcendance of the human condition.[2] Today, its ranks are divided between cosmic utopians who see in the potential powers of science and technology the total liberation of mankind, and the catastrophists who see the limits of the scientific imperative as having been met. One basic understanding unites both factions, however; namely, that "the future is in doubt today as never before."[3]

This doubt has, however, served as an impetus for venture capitalists to intermingle science and business and be willing to take risks.[4] Indeed, the short history of biotechnology appears to be repeating the very pattern of innovative capitalism seen typically in the United

States.[5] It has been predicted that by 1990 this new industry will have a capitalization of anywhere from $3 to $5 billion;[6] and by the year 2000 achieve a growth of $40 billion.[7] It was reported recently that approximately one out of every ten biotechnology firms in the United States is in the Baltimore, Maryland–Washington, D.C. region that in turn means as many as 130 companies employing 3250 of the nation's 40,000 biotech workers are in this area. Regional predictions of the biotechnology work force call for an increase to approximately 15,600 by the year 1995.[8]

THE PROCESS OF ALTERATION

Essentially, the genetic composition of future generations may be altered in three ways: environmental strategies and policies, eugenic programs, and genetic engineering.[9] Regarding environmental changes, not only do discoveries in medicine, but the institution of a National Health Service, efforts to develop programs for poverty relief, agricultural changes and alterations in the tax position of large families all serve to alter the selective pressure on genes. Indeed, it is not difficult to think of any social change that does not make some difference to those who survive or who are born.[10]

Eugenic policies designed as such to alter the breeding patterns or survival patterns of people with differing genes may also be termed "environmental."[11] The third method to change the genetic composition of future generations is to be found in the use of genetic engineering: using enzymes to add or subtract from a stretch of DNA.[12]

In the 1970s, techniques were discovered by which DNA (deoxyribonucleic acid) molecules could be cut into pieces in a controlled manner and then recombined in new ways, by transferring the cells of another species and fitted into the chromosomes there, or by a process where pieces of DNA molecules are taken from one species and then combined with pieces from another species. Refinement of recombinant DNA experiments could produce easily genes in organisms that had never had them previously or, for that matter, even genes that heretofore had never been in existence *anywhere*.[13] Thus it is seen that the dimension of the combined influences of biology, biochemistry, and genetic engineering provides a monumental foundation for biotechnology and at the same time provides a new and exciting way of organizing nature.[14]

Fears and Concerns

The idea of genetic engineering being used to produce creatures combining characteristics of more than one species—and especially when one is human—promotes revulsion in some people. There are two central concerns here: the fear that the product would be viewed as an outcast and not accepted as a member by the species from which it came, and the fear that inappropriate parts of different species could be combined. More specifically, it is commonly thought that a creature with a fair amount of human mentality would be unable to express much if he had a body derived from a wolf or cat.[15]

Cyborgs and Beyond

A relatively new word, "cyborgization," has been coined to define the integration of the biological systems

of man's body with mechanical systems.[16] In developing
artificial limbs and artificial organs, medical science has
created the first generation of cyborgs or cybernetic organisms.[17] As opposed to genetic engineering that seeks
to achieve its goal by direct manipulation of the genes,
cyborgization opens up possibilities in human engineering in dramatically different ways. Thus, if a man
could be equipped to link up to *machines*, he could ef-
fectively acquire a whole range of extended selves by
being fitted into machines of many different kinds. Already, some individuals suffering from kidney disease
have been linked up with machines through the use of
plastic sockets on parts of their bodies which allows
them to plug themselves straight into a kidney machine.[18]

Improvements in human capability might be more
readily achieved through cyborgization than by efforts
to transform human egg cells. Accordingly, if individuals wished to be equipped with new senses, efforts
could be turned to mechanical implants and artificial
synaptic interfaces rather than try to build in new sen-
sory equipment by manipulating the genes, or by mod-
ifying the bodies of adults. Interestingly, a radio receiver
that can be attached directly to the auditory nerves—
powered either by durable batteries or by the body's
own biological systems—is presently within the range
of feasibility.[19]

Manufactured Processes

The biotechnology revolution is not only developing new sources of power to alter creatures, including
humans, but promoting new processes that enable the

manufacture of all the materials that nature has used throughout those billions of years that the Earth has been in existence. Ultimately, the power will even be found to design and produce new materials that natural selection never managed to invent.[20]

The first applications of organic technology to biotechnology are being found in reproduction of substances already known in nature.

> The polysaccharide chitin, used by insects and fungi to supply the framework of their bodies, might have as much potential as wood (which is mostly made of other polysaccharides, cellulose and legnin). Ivory from elephants' tusks is at present a scarce and precious commodity, but biotechnology might be able to produce it in much more considerable blocks in the fairly new future. Genetic engineers have already cloned the gene for one of the four components of tooth enamel. Selective breeding has produced silk works which are marvelously prolific in spring that they wrap around their cocoon, but biotechnology holds out the prospect of cutting out those living middlemen in much the same way as it promises.[21]

While the range of products of the new biotechnology is virtually limitless, their marketing remains different. For example, if a new hybrid corn seed was developed through efforts of genetic engineering, how is it to be sold to the 20,000 farmers in the country against a corn seed that such a major seed company as Pioneer sells currently?[22] Similarly, how can a new prescription drug be marketed to thousands of physicians unless a marketing force exists to make visits or calls? Producing the products of biotechnology cannot even be undertaken unless manufacturing facilities exist and a structure is in place to meet all the government regulatory controls.[23]

The Genome Project

A new major proposal to decipher all the 100,000 genes in the human body, known collectively as the genome, has been started. It has been described as "the holy grail of biology" because it could well provide a detailed base of knowledge concerning the identity and eventually the function of all human genes, that in turn would provide physicians with a genetic blueprint, of sorts, of normal development, together with a map of the causes of a wide range of inherited diseases. Congress appropriated, in December 1987, $17.4 million for the National Institutes of Health, and $11 million for the Department of Energy to commence development of the technologies that will be needed ultimately for the genome project.[24] The National Academy of Sciences urged some $200 million a year in new appropriations to start the project, that over 15 years could reach $3 billion.[25]

An Innovative Application

In May 1987, the United States Patent and Trademark Office announced that it "considers non-naturally occurring nonhuman multi-cellular living organisms, including animals to be patentable subject matter."[26] Although viewed by the Patent Office as but an effort to keep pace with the startling new advances in biotechnology, and thereby encourage innovation and not determine its ethical implications, others—such as animal rights advocates—were concerned that animals were being considered as products and not sentient beings.[27] Some feared also that the new policy would enable a select number of biotechnology companies to dominate the livestock industry, thereby eliminating small independent breeders and seeking to eliminate genetic diversity among farm animals,[28] since with patents the central issue becomes who either owns or is in control of breeding livestock.[29]

Theologians quarreled with the Patent Office policy because it not only equated heaven-made creatures with manufactured goods of the market place, but took a giant step on the slippery slope that would lead to the patenting of genetically altered human beings and man's full assumption of Godlike powers. The clear specification of the policy that its application was only for "nonhuman life" was of no reassurance here.[30]

Informed members of the scientific community saw the Patent Office as merely continuing the reasonable exploitation of nature. As a director of Ohio University's Animal Biotechnology Center in Athens, Ohio, said succinctly: "A pig is a pig, and a cow is a cow. You merely enhance certain aspects of it."[31]

It is expected that the near future of biotechnology will give rise to work in laboratories in the United States where virus and bacteria genes will be transferred to plants in an effort to enable them to produce their own particular insecticides or fertilizers. After field testing, these "transgenic" plants will in turn be used by farmers in the place of conventional crop varieties.[32] Further successful research will be undertaken that manipulates the primordial cells producing sperm and eggs to enable breeders in turn to determine the sex and other preferred characteristics of their animals; and routine gene transplants from one species to another will be accomplished routinely.[33]

Already the Federal Department of Agriculture operating from its research center in Beltsville, Maryland, has produced a rust-colored "transgenic" pig that was bred with the growth hormone of a cow. Engineered with the idea of achieving less fat, the pig has met this scientific purpose. But, sadly, it also suffers from severe arthritis and thus has difficulty walking and has crossed eyes as well.[34] A policy group opposing genetic engineering, the Foundation of Economic Trends, together with the Humane Society, unsuccessfully maintained a legal action against the efforts of the government to halt the research that produced this particular boar's sire. The essence of their claim was that research of this nature not only was cruel and violated animal dignities, but would also have significant social and economic repercussions, in that more expensive animals would in turn cause severe market dislocations in the farm economy.[35]

Since no catastrophic events have followed the decision by the United States Supreme Court in 1980 to allow new forms of laboratory life to be eligible for patenting,[36] none are likewise expected from this new policy of the United States Patent and Trademark Office. The debates will nonetheless continue over the long-range effects of genetic engineering and its ethical constraints. Yet they will have little real value in halting the momentum of scientific inquiry, experimentation, and advancement of biotechnology.

As the director of The Hastings Center in New York, an organization devoted to the continuing ethical study of the advances of the new biological technologies on society, stated, "It's very hard to sustain a great deal of worry about these things when, after 10 years of pretty constant interest and attention, there have been no untoward events."[37]

SCIENTIFIC FREEDOMS VERSUS SOCIAL RESPONSIBILITIES

On November 20, 1973, Julius Stone presented the tenth annual Mooers Lecture, entitled, "Knowledge, Survival, and The Duties of Science," at American University in Washington, D.C.[38] The central question and theses which he propounded then could, and indeed should be raised anew today, for they form the very core of the province and function of law, science, and medicine in our brave new world of today and tomorrow; and they point also to the leeway of choice and patterns of discourse that exist in grappling with this central issue and possibly forging a consensus opinion for a subsequent course of action.

Dr. Stone admonished us to be aware of the "genetic revolution" where people would be created in test tubes and molecular "monsters" would be released into the atmosphere.[39] He proceeded to caution that "the liberty to extend knowledge is not absolute," but must be limited when it is in conflict with other values.[40] He posited the central question of his inquiry as being a study of the extent to which scientists "have a moral duty to consider, along with others of competent knowledge, whether a line of inquiry should be desisted from as soon as it becomes clear that it is likely to bring about a mankind-endangering situation, which no one has any foreseeable capacity to handle. . . ."[41]

Although it is debatable whether it is totally impossible to reverse the process of discovery, Stone suggested that a particular scientist might well have a moral duty "not to contribute by *his* work to the certainty or speed of its arrival."[42] He acknowledged that the essential role of the scientist is to advance knowledge and that compromises should not be freely undertaken that limit his inherent or fundamental freedom to so act and, further, "that whether knowledge is put to good or evil use is a matter for society generally, and not for scientists."[43]

From this, Professor Stone shaped his thesis to state accordingly that: "scientists have a duty to exercise self-restraint in pressing further those scientific activities which manifest" a likelihood that they will result in "limit-situations" or, in other words, "dangers of cataclysmic physical or psychological proportions for mankind as a whole," and specifically where the particular scientist in question is actually "aware of this likelihood as a proximate outcome" of his own work.[44] He stressed the point that the scientific duty of restraint should only be imposed when the scientist is "clearly able to foresee that the particular line of work is leading to a scale of dangers" that would constitute a "limit-situation."[45]

Thus Stone delimits the scope of scientific inquiry to a very narrow but admittedly crucial range.[46] He observed that his essential inquiry is "not whether scientists should cease all activity that might lead to *any dangers*, much less that they should always be able to foresee *all* consequences."[47] Rather, it is tied to "whether they should not desist from activities likely to lead to dangers cataclysmic for mankind, *and* against which no protection seems possible, *from the moment at which they can already foresee these dangers.*"[48]

He admitted that the criteria he submitted for determining restraints on scientific inquiry lacked "precision"; but he explained that, "the indeterminacies leaned in favor of the traditional scientific freedom of investigation" and that "no duty of restraint" arose unless the scientist was able to foresee for himself the magnitude of the dangers of his research. He contended further that even though elements of indeterminacy were present within the criteria which he postulated, they "give guidance to all concerned" in that they not only indicate "the relevant orders of magnitude and imminence but also the nature of the substantive values threatened."[49] More specifically, Professor Stone was concerned with two orders of such values: one that embraces the limits of physical integrity and the sanctity of human life together with mankind's survival in general and the second one concerned as such with the dangers arising from "scientific advances to human individuality, in the sense of the autonomy of the human will and sensibilities presupposed by our notions of freedom."[50]

Stone expressed his grave reservation about the feasibility of in vitro fertilization as well as genetic surgery and engineering.[51] Although he recognized barren marriages could be resolved by the new noncoital techniques for reproduction and, further, that genetic engineering could alleviate genetic-borne disease and disability, he "would not admit that relief afforded for such cases (admirable in itself though it might be) could even begin to tip the scales against the formidable dangers to a liberty-based society to which test-tube birth or any analogue of this would open the way."[52]

There can be little quarrel with Dr. Stone's idea of social responsibility in scientific inquiry and investigation. I find myself, however, in respectful dissent to his concern regarding the dangers of research into the fields of the noncoital reproductive sciences. So long as procreation continues to remain the central driving force in a marital relationship and the family the very core of a progressive society, efforts will be undertaken to expand the period of fecundity and combat infertility itself. Genetic planning and eugenic programming are more rational and humane alternatives to population regulation than death by famine and war.

Human Rights and the New Technology

On the sophisticated continents such as America, Europe, and Australia, the pervasive attitude has been, until quite recently, quite supportive of scientific inquiry and discovery, for it was believed that this action was not only of overwhelming benefit to society, but an essential attribute of human achievement and progress.[53] Subsequent agonizing reflections on the horrors of World War I and World War II and the all too frequent limited conflicts since 1945, together sometimes with overly emotional concerns regarding the full potential for annihilating mankind, have witnessed a new and increasingly pessimistic temperament concerning scientific advancement. Indeed, it has been recognized that "not all science is good for humanity."[54]

The importance of human rights and their need to be recognized in the era of the New Biology was underscored by initial efforts at the United Nations in the 1960s.[55] But before that activity, the 1948 Universal Declaration of Human Rights' guarantees of "human dignity" written in Articles 1, 5, 6, and 29(1)[56] established eloquent reminders of the need for the advances of biotechnology and genetic engineering to be tied to a basic understanding of and respect for fundamental human rights.[57] Indeed, what is needed now is a new human rights debate within not only the legal community, but also with scientists and technologists; a debate that would consider anew the extent to which both the traditional and the redefined rights of humanity are challenged or, as the case may be, complemented by the plethora of medical, legal, scientific, and technological considerations of the brave new world that is already here. Justice Michael D. Kirby succinctly summarizes the issue when he says:

> If lawyers are to continue to play a relevant part in the human rights debate of the future, they must become more aware of scientific and technological advances. Otherwise, they will increasingly lack understanding of the questions to be asked, let alone answers to be given.[58]

CONCLUSION

Law has all too often been found, in the words of Justice Windeyer, to be "marching with medicine but in the rear and limping a little."[59] Law, science, and medicine must become full and not limited partners and march in unison as they approach the task of assuring the primary goal of society both today and tomorrow that all citizens have an equal opportunity to achieve their maximum potential within the economic and biotechnological marketplaces, have their physical suffering minimized, and their spiritual tranquility assured.[60]

Chapter 2

Law, Science, and the New Biology

INTRODUCTION

Today, scientific work is less a basic expression of the "ancient aristocratic ethos of the love of knowledge" than a mere job to be done—by entrepreneurs, employees, or others who have independent funding.[1] In 1980, Genentech—a San Francisco based biotechnology company—was the first such company to issue shares on the over-the-counter market. Among its products are a hormone capable of stimulating human growth, mass-produced human insulin which would allow a substantial reduction in cost of the treatment of diabetes, and interferon which may prove to be the long awaited "miracle" drug to combat cancer. In 1984, the Office of Technology Assessment estimated conservatively that some 225 firms are engaged in "commercializing biotechnology."[2]

It has been asserted that patenting new forms of

life, as sanctioned by the United States Supreme Court,[3] will be guided by short-term profit motives rather than sound philosophical principles.[4] However, scientific knowledge is not, in and of itself, an absolute end. The thrust and purpose of patenting new life forms are basically technological and are essentially political. Because the etiology of new life forms is political, both its costs and its benefits are, of necessity, of public interest and concern.[5]

Pure scientific inquiry does not produce an economic exploitation of nature; only man's use of the truths of scientific inquiry does. The methodological style of nature science seeks to demonstrate causal relations among events. Thus, the laws of science state that whenever X occurs or varies in a particular way, Y will similarly occur or vary in a particular way. This phenomenon has been aptly termed "a formula for action." Its practical application awaits only an individual's decision that it might be economically advantageous to try to mobilize X's to produce Y's.[6] Science promises truth, not peace of mind.[7] Yet liberty to extend knowledge is never to be regarded as absolute, but rather undergoes limitation when it conflicts with other values.[8]

The purpose of this essay will be to explore the parameters of the scientific imperative to explore truth. The scope of this inquiry is shaped in part by the United States patent laws and administrative interpretations and, more specifically, by the United States Supreme Court in its momentous holding allowing new forms of life created in a laboratory to be patented. The ultimate purpose of this investigation, then, is to refute the arrogance of power theory expressed as being implicit in the current studies of the vast potential for the positive achievement of good through harnessing the New Biology. Thus, I intend to demonstrate that what has been dismissed as but a magnificent obsession for power, profits, and immortality has in truth a far more intrinsic potential for good and reward for the scientific community and the greater world community.

Improvement of man's genetic endowment by striving for positive propagation of those with a superior genetic makeup or, conversely, delimitation of those with negative genetic inheritance has always been a primary concern in the field of genetics.[9] Accordingly, if in some way the quality of life can be either advanced or strenthened by and through the use and the application of law as it relates to and indeed shapes genetic policies, then such actions must be pursued.

ALTERING HUMAN EVOLUTION

Today, man is in a position not only to alter the social and environmental conditions of the universe, but also to change his very essence.[10] The mythology of the Minotaur and the Centaur, half man and half animal, may well become the reality of the twenty-first century. Indeed, modern medicine is presently not only attempting to create man-animal combinations, but also man-machine combinations or cyborgs.[11] Plastic arteries, artificial hearts, electrically controlled artificial limbs, and pacemakers highlight the achievements of modern science to replace diseased or worn-out parts of the human body.[12]

Efforts to construct or engineer biologically functional bacterial plasmids in vitro exemplify the relatively new technology of recombinant DNA.[13] Regarded as the most significant step in the field of genetics since 1953, research in this technology will facilitate identification of every one of the 100,000 genes in the human cell.

Armed with this information, efforts could be directed toward replacing defective genes with healthy ones. Thus the hope is that in making such replacements, genetic diseases such as hemophilia and sickle-cell anemia could be conquered.[14] Indeed, the plentitude of new products of nature that could substantially improve the human condition is staggering to the imagination.

The National Institutes of Health have taken a conservative view of the limits of safety review required by those institutions receiving federal grant monies to experiment in DNA. In 1980, however, 200 representatives from the scientific community called upon NIH to loosen the restriction on gene-splitting experiments conducted in the United States. The scientists expressed the growing agreement that DNA research carries with it fewer risks than had once been thought.[15]

The central question that arises in relation to the current scientific advances is whether genetic engineering should be promoted and encouraged as a basic recognition of the freedom of scientific inquiry and right of privacy. Significant potential dangers are present in conjunction with the almost limitless opportunity for scientific advancement within the technology of recombinant DNA, commonly referred to as genetic engineering. The fear that the proverbial "mad scientist," working independently or with an enemy foreign power, could isolate and then proceed to duplicate a cancer organism and place it—possibly—in public water supplies is not easily dismissed. Acts of thoughtless negligence in a laboratory could result in the "escape" of a deadly microbe which in turn could give rise to a "parade of horribles." Chance occurrence is always inherent in any scientific intervention.[16] When the chance of harmful accident is calculated, the primary consideration is whether the merit of the intervention justifies beginning or continuing the experiment.[17]

Genetic engineering, viewed as an instrument to revolutionize, limits the effect of natural selection and replaces it with programmed decision making. Programmed decision making in turn serves to facilitate rational thinking rather than impede it. Is it shameful to acknowledge that man has the capability to be in control of himself? The lack of control over the centuries spawned a type of "evolutionary wisdom" which resulted in the bubonic plague, smallpox, yellow fever, typhoid, diabetes, and cancer. Today the quest for maximum efficient utilization of biological and medical knowledge represents one of the tenets of the so-called "evolutionary wisdom."[18]

A number of post-Darwinians in the scientific community assert that there is no wisdom in evolution, only chance occurrence. Few if any would be willing to accept unconditionally all that nature bestows, particularly disease. Consequently, science finds itself in the position of trying to both influence and, in many cases, control the process of evolution. Some would go so far as to suggest that dangerous knowledge is never half as dangerous as dangerous ignorance.[19]

The sanctity of creation and the fundamental right of privacy in procreation, which is an acknowledged basic or fundamental freedom, may be altered by compelling state interests.[20] Is there a more compelling state interest than the desire to stop a "chromosomal lottery" that saddles the economy each year with 4 million Americans born with diabetes or 50,000 born with discernible genetic diseases?[21] State interests in minimizing human suffering and maximizing the social good should be properly validated.[22]

Opponents of unrestricted genetic research specifically attack its proponents as being both scientifically and socially irresponsible and the ultimate promoters of environmental disaster.[23] They suggest that nature has developed strong barriers against genetic interchanges between species and that extreme caution ought to be used during experimentation in this area.[24] Others argue that mankind's genetic inheritance is its greatest and most indispensable treasure which must be protected and guaranteed at any cost. These opponents submit that the evolutionary wisdom of the ages must not be irreversibly threatened or abridged in order to satisfy the ambition and professional curiosity of some members of the scientific community.

Autonomy, self-determination, and a basic sense of freedom must be tempered by logic, objectivity, and a disinterested search for knowledge, a search that may result in the minimizing of human suffering and maximizing of social good.[25] But what *is* the social good in this question? It is suggested that the social good within this context could be equated with an economic policy that lessens the financial burden on citizens and supports and maintains genetically defective citizens. The wisest policy is, by consensus, that which promotes a good—social, economic, or other—for the greatest number. Thus, human need and well-being shape the degree of positive good resulting from one policy as opposed to another.[26] Alternatively, a determination could be made in order to structure what is right or wrong, good or evil, according to whether the consequences of an act or public policy add to or detract from the aggregate human well-being.[27]

Ultimately, the decision for or against a policy is going to be tied to development and maintenance of an a priori standard or ethics (where, in theory, a balancing occurred before the standard was set) or to a situation ethic by which the consequences, pro and con, equities or inequities, of each proposed action will be carefully weighed and a conclusion with an ethical posture or structure of a standard of modus operandi[28] will be reached.

ENCOURAGING EXPERIMENTATION

Recognizing that a sustained level of progress for society would depend upon a continuing standard of technological evolution as well as individual technological contributions of exceptional merit and benefit, the Founding Fathers endeavored to codify this attitude within the Constitution itself. By structuring a system of checks and balances within the Constitution that would promote both perspectives, contributions that were truly exceptional could be promoted by grant of a limited monopolization as authorized by the Patent Clause.[29] However, the grant of limited monopolization was intended to be consistent with the guarantees of the Fifth and the Fourteenth Amendments, that recognize the right of all citizens to develop their individual skills in pursuit of a trade or calling, and thus establish the right itself as an inalienable property right.[30]

The recorded history of efforts to legitimize monopolies for patents of unworthy inventions is long. To its credit the United States Supreme Court has thwarted these efforts and has thus recognized and enforced the constitutional mandate to allow the unfettered growth and natural evolution of technology.[31]

On June 16, 1980, by a 5 to 4 vote, the United States Supreme Court decided that new forms of laboratory life were eligible for patents.[32] The decision may be regarded as a ratification of some of the accomplishments of the "biological revolution" which has allowed a broader understanding of life and promoted a greater ability to manipulate various forms. However, both the majority opinion and the dissent stressed that they address only the question of whether the current patent laws evinced a congressional intent to deny patents to those inventions determined to be alive.[33] More particularly, the Court chose to tie itself to the United States Code section which provides: "Whoever invents or discovers any new and useful process, machine, manufacture, or composition of matter, or any new and useful improvement thereof, may obtain a patent therefor, subject to the conditions and requirements of this title."[34] Out of this statute emerged the issue of whether a manufactured microorganism constituted a "manufacture" or "composition of matter" within the meaning of the statute.[35]

Dr. Ananda M. Chakrabarty, a microbiologist employed by the General Electric Corporation, engaged in research in which he succeeded in manufacturing a new microorganism, not found in nature, which is effective in breaking up oil spills. This genetically engineered string of *Pseudomonas* is made by combining (or crossbreeding) four strains of oil-eating bacteria into one man-made scavenging microorganism which combines the beneficial properties of each of its four parent bacteria. Each of the four strains digests particular hydrocarbons in a mixture of oil and water—such as is found in petroleum spills. Useful by-products of water, carbon dioxide, and a bacterial protein which is nutritious to inhabitants of the ocean, remain. Dr. Chakrabarty demonstrated that this manufactured "superstrain" is much more efficient in digesting oil than a mixture of the four individual bacteria. Another advantage is that this microorganism, if it "escaped," would not be able to thrive in gas tanks or in the oil fields of the earth and wreak uncontrolled environmental havoc on the ecosphere.[36] The Chakrabarty bacterium had already been granted a patent in Britain, which had followed several European nations in recognizing both plants and animals as patentable.[37]

The patent application of Chakrabarty and General Electric was for a manufactured microorganism product not found in nature, as well as for a process of using the microorganism on a carrier to digest oil spilled in water. The United States Patent Office rejected the product claim, but allowed a portion of the process claim. The rationale for rejection of the product claim was that a living organism—a naturally occurring product of nature—as this was determined to be, was not within the classes of subject matter which are patentable. The Patent Office reached this conclusion because there was no mention of such a class in the controlling statute or in the statute's legislative history. This decision was upheld by the Patent Office Board of Appeals, but the United States Court of Customs and Patent Appeals reversed, and the Patent and Trademark Office appealed to the United States Supreme Court.[38]

In the past, the Patent Office has included living things within the statutory subject matter. For example, in 1873, United States Patent No. 141,072 was issued to Louis Pasteur. Claim Two of the patent application reads: "Yeast, free from organic germs of disease, as an article of manufacture."[39] There are other examples, in other patents, of claims having been granted for viruses and cultures.[40]

Today, there are more than 100 patent applications related to products of genetic engineering.[41] Chakrabarty sets the pace for a wide variety of new "man-made organisms" which can facilitate socially desirable processes such as growing wheat in arid lands, leeching ores to assist mining companies in reaching remote parts of the earth, and producing a "bug" that will ferment cornstarch or corn syrup into ethanol, an alcohol used in both whiskey and gasohol. There is also a patent application for a bacterium that metabolizes ethylene into ethylene glycol (antifreeze).[42]

As noted previously, the major thrust of the decision of the United States Supreme Court in Chakrabarty is tied to the interpretation of the term "manufacture" as it appears in the Federal Patent Code. Observing that Thomas Jefferson's Patent Act of 1793 stressed its coverage to "any new and useful art, machine, manufacture, or composition of matter, or any new or useful improvement [thereof]," Chief Justice Burger, writing for the majority, defined manufacture as "the production of articles for use from raw materials prepared by giving to these new materials new forms, qualities, properties, or combinations whether by hand labor or by machinery."[43] Citing approving precedent defining "composition of matter" as including "all compositions of two or more substances . . . all composite articles, whether they be the results of chemical union, or of mechanical mixture, or whethr they be gases, fluids, powders, or solids," the Chief Justice concluded that the Chakrabarty microorganism qualifies as being within patentable subject matter.[44] The claim is particularly forceful since it is for a product of human ingenuity which is non-natural in its occurrence.[45]

In response to the argument that microorganisms cannot be patentable without express congressional authorization, the Chief Justice declared that Congress had already defined what was patentable subject matter in Section 101 of the Act, and that it was for the courts to define that provision. Finding no ambiguity in the statutory provisions and stressing the broad constitutional and statutory goal of promoting "the Progress of Science and the useful Arts," Chief Justice Burger adhered to his position that the definition the Court gives to Section 101 is consistent with the goals of the Act.[46]

The Court declined to acknowledge the "grave risks" or the "gruesome parade of horribles" which the Patent Office argued that the Court should weigh in deciding whether the Chakrabarty invention is patentable.[47] Although acknowledging that "genetic research and related technological developments may spread pollution and disease, that it may result in a loss of genetic diversity, and that its practice may tend to depreciate the value of life," the Court concluded that neither the grant nor the denial of patents on microorganisms will end advance in genetic research nor "deter the scientific mind from probing into the unknown any more than Canute could command the tides."[48] The Court stated unequivocally that scientific arguments against advancements in this field are matters of "high policy" that should be considered by the legislative process which balances and places in proper perspective the various competing values and interests of all interested parties.[49] The Chief Justice concluded by noting that if the Court has misconstrued the provisions of Section 101, all that Congress needed to do was to amend the statute so as to exclude from the protection of the patent laws organisms which are produced by genetic engineering.[50]

Despite the Court's disclaimer that its action was purely constructive in nature—merely an interpretation of a statutory mandate—it did attempt to validate a new national policy. While invoking the Jeffersonian concept of ingenuity in patent creativeness, it came down foursquare on a policy encouraging experimentation into the

New Biology despite the possible risk to mankind. Thus, while disclaiming the application of a balancing test, it—in effect—performed one. It correctly decided that the utility of the good that will flow from research and experimentation into the varied fields of the New Biology far outweighs the potential harm accruing as a consequence of such undertaking. This is an eminently fair and reasonable position.

CONCLUSIONS: TOWARD A STANDARD OF REASONABLENESS

The Supreme Court's actions in Chakrabarty give private corporations the incentive to invest in further research into the fields of biochemistry, genetics, and eugenics. This incentive and the anticipated result therefrom satisfy the constitutional objective of early disclosure which in turn expands the public domain of knowledge in these fields. There can be little doubt that patentability of microorganisms is "Progress of the Useful Arts."

Man's dehumanization and depersonalization will not be fostered as a consequence of the continued quest for mastery of the genetic code. Attendant on the freedom to undertake research into the exciting and fertile frontiers of the New Biology is a coexistent responsibility to pursue the work in a reasonable, rational manner. Pursuing the New Biology in such a manner requires adequate attention to the safety factor in all aspects of the experimentation.[51] The undesirable elements of a Brave New World can be tempered only when knowledge is pursued with the purpose of establishing the truth and integrity of the question, issue, or process.[52] The vast potentials for advancing society and ridding it of a multitude of its present ills is an obvious good which must be pursued steadily. Little sustaining harm can result from a reasonable pursuit of truth and knowledge; for, indeed, truth and knowledge are the basic interstices in any balancing test.[53] If actions are undertaken and performed with the goal of minimizing human suffering and maximizing the social good, then the noble integrity of evolution and genetic progress will be preserved.

Chapter 3

Medical, Legal, and Ethical Conundrums at the Edge of Life

INTRODUCTION

Each year, approximately 30,000 genetically handicapped at-risk infants are born in the United States.[1] In Australia alone, it has been estimated by the new South Wales Health Commission that approximately $500,000 will be spent during an average lifetime for one institutionalized person with a genetic abnormality.[2]

Initial decisions regarding the administration of care—heroic and extraordinary, or nursing and ordinary—to the defective newborn present a number of vexatious issues. The purpose of this essay is to explore a number of those very issues. Toward the achievement of this end, it has been developed into six parts dealing with: the establishment of a working definition, legislatively and administratively, of handicapped individuals; the British and the American judicial postures regarding the matter; the medical profession's response;

triage as a viable principle for the allocation of scarce neonatal support systems; the expanded role of ethics review committees; and the development of a construct for effecting humane decision making.

Both the thesis and the major conclusion of this chapter are identical and state—very simply—that no matter what physicians, lawyers, judges, social workers, philosophers, or ethicists posit re the structuring the validation of treatment or nontreatment of handicapped newborns, the final reality of decision making is tied to a complex balancing test of weighing economic costs of nonmaintenance against the social benefits of maintenance. Stated another way, decisions of this scope and dimension are reached by balancing the gravity of the economic harm that will accrue in a particular case of maintenance against the utility of the social good that will occur for nonmaintenance.

DEFINING THE HANDICAPPED

Since the mid-1960s, neonatology has developed as a subspecialty in pediatrics and neonatal intensive care units have emerged.[3] These new units make it possible to save the lives of newborns who previously would have died. While this technological progress must be recognized with excitement, its continued development has intensified the critical moral dilemmas now faced by health care providers. Sophisticated neonatal care is not only costly, but scarce, with demand often exceeding the supply of beds, equipment, and personnel. Physicians and hospitals thus are often forced to choose *which* newborns receive intensive care and which do not.[4] Decisions of this nature in turn involve a plethora of medical, moral, legal, and economic problems, with the central issue being how best to distribute scarce neonatal intensive care resources.

As observed, costs are of considerable importance in this sphere of decision making. The United States Department of Health and Human Services reported the Nation's health bill shot up 15.1% in 1981, outpacing the 8.9% inflation rate for the year. Overall medical costs, both public and private, rose to 287 billion dollars. This sum represented a record of 9.8% of the gross national product averaging out at a cost of $1225 for each American.[5] Over the past 10 years in the United States, hospital care expenditures quadrupled to $118 billion dollars, with the costs of physicians' services more than tripling to 54.8 billion dollars.[6]

A Legislative Approach

The Rehabilitation Act of 1973 enacted by the United States Congress[7] and modeled in large part on the Civil Rights Act of 1964[8] and the Education Amendments of 1972[9] sought, through Section 504, to enunciate a federal policy that would mandate nondiscrimination against handicapped individuals who participated in programs that received federal financial assistance.[10] The Congress chose to define "handicapped individual" in vocational terms as:

> . . . any individual who (A) has a physical or mental disability which for such individual constitutes or results in a substantial handicap to employment and (B) can reasonably be expected to benefit in terms of employability from vocational rehabilitation services provided. . . .[11]

In 1974, the Act was amended, the concept of "handicapped individual" redefined, and Section 504 by implication extended to all government programs receiving federal financial assistance. The new expanded definition of "handicapped individual" was held to apply to:

> . . . any person who (A) has a physical or mental impairment which substantially limits one or more of such person's major life activities, (B) has a record of an impairment, or (C) is regarded as having such an impairment.[12]

The Rehabilitation Act has involved the full spectrum of legal interests from employment discrimination[13] and discrimination in education[14] to discrimination as a result of inaccessability.[15]

It remained for President Ronald Reagan in 1982 to interpret and apply Section 504 to treatment decisions regarding defective newborns.[16] Accordingly, an Interim Final Administrative Rule was promulgated by the United States Department of Health and Human Services which modified presently existing regulations in the Department by including provisions for the medical care of defective newborns.[17] Specifically, the Department advised all federally assisted hospitals (which number over 7000) that their failure to comply with the expanded definition of Section 504 as applied specifically to newborns could well subject those hospitals to termination of their federal financial support.[18]

The novelty of expanding Section 504 as a basis for preventing discrimination has been fraught with difficulty and misunderstanding since its early operation. In fact not until February 23, 1984, was it finally determined by a federal court that Section 504 of the 1973 Rehabilitation Act was *not* applicable to treatment decisions involving defective newborn infants.[19] Prior to this decision, the role and application of Section 504 actions was ill-defined. President Reagan's 1982 actions regarding this Section were thwarted by a federal district court decision in 1983 in the case of American Academy of Pediatrics v. Heckler,[20] which held the Rule promulgated by the United States Department of Health and Human Services, consistent with the Reagan Directive, was arbitrary and capricious.[21]

The impact of the February 1984 judicial opinion in United States v. University Hospital, State University of New York at Stony Brook, et al. was delayed. For on January 12, 1984, new revised rules and regulations were issued concerning nondiscrimination on the basis of handicap relating to health care for handicapped infants based upon what was understood as the enabling authority of Section 504.[22] On March 12, 1984, several medical groups including the American Medical Association, and the American Hospital Association, filed an action in the United States District Court for the Southern District of New York which sought to prohibit the Department of Health and Human Services from implementing the new, revised rules issued in January; and on May 23, 1984, the Court ruled that the regulations were issued without proper statutory authority.[23]

While the Justice Department perfected its appeal of this case, the 98th Congress considered proposals in the House of Representatives[24] and in the Senate[25] to amend provisions of the Child Abuse Prevention and Treatment Act and the Adoption Reform Act of 1978 designed to make the withholding of medical treatment from handicapped babies with life-threatening conditions the basis for an action of child neglect and abuse.[26] Public Law 98-457 was enacted subsequently on October 9, 1984, and entitled The Child Abuse Amendments of 1984.[27] The Secretary of Health and Human Services promulgated new regulations designed to implement the law on December 10, 1984.[28]

While the basic policy of these new model guidelines is to prevent the withholding of medically indicated treatment from disabled infants with life-threatening conditions, the guidelines specifically do not apply and thus do not mandate a course of treatment where a physician's reasonable judgment is that: the infant in question is chronically and irreversibly comatose; where treatment would merely defer death, not be effective in ameliorating or correcting all of the infant's life-threatening conditions, or otherwise be futile in terms of the survival of the infant or, finally, such treatment would be "virtually futile in terms of the survival of the infant and the treatment itself under such circumstance would be inhumane."[29] Pursuant to these new proposed guidelines, hospitals are encouraged to establish Infant Care Review Committees (ICRC's).[30] It is especially recommended by the Department of Health and Human Services that ICRC's be established in tertiary level neonatal care units both to effect and thereby implement the guidelines.[31]

The States Respond

In the U.S., three states have enacted specific legislation concerning emergency care and treatment of handicapped at risk infants.[32] Under the Arkansas legislation, a parent or legal guardian is allowed to sign a declaration on the behalf of the infant or incompetent in question prohibiting the use of extraordinary means of care and treatment in order to prolong the at-risk individual.[33] It would thus appear on the basis of this legislation that the parents or guardian may make an autonomous and indeed unreviewable choice to allow an infant's death so long as the level of care necessary to prolong life is characterized as extraordinary. Generally, the actual life-threatening condition that a disabled infant suffers—an intestinal or esophageal blockage—may be remedied by what is considered to be relatively routine surgical interventions. It would thus remain to be seen whether this controlling Arkansas statute would be in fact construed as applying to this group of cases. Extraordinary means are, predictably, nowhere defined in the statute.

New Mexico has a similar statute which allows parents or guardians to direct the withholding of care from terminally ill minors.[34] Yet this statute is circumscribed to a greater degree than the Arkansas legislation in that care may only be withheld from individuals classified as "terminally ill"; which is defined as those who, regardless of the use of extraordinary care, will nonetheless die as a result of their illness.[35] This obviously would not cover those infants who would survive in the event standard corrective surgery were to be performed. Yet this law would probably be applicable in cases of hydrocephaly or anencephaly, since infants suffering from conditions of this nature will in all likelihood die eventually as a direct result of the condition itself. Unlike the Arkansas statute regarding the nonreviewability of parental decisions not to treat, under the New Mexico legislation, provision is made for appointment of a guardian *ad litem* to represent the interest of the at-risk infant.[36] It remains unclear, however, whether the guardian will enjoy an effective role in the proceeding, since no requirement is imposed which mandates that the decision ultimately made regarding the continuance or noncontinuance of treatment be in the best interest of the infant. Consequently, on the face of the statute, it would thus appear that the parental decision must stand if all legal formalities are met.

In marked contrast to these two legislative pronouncements, the statute in Indiana declares that once a fetus is born alive it must be treated thereafter as a person regardless of its physical or mental condition.[37] The statute declares that "failure to take all reasonable steps, in keeping with good medical practice, to preserve the life and health of said live born person shall subject the responsible persons to Indiana laws governing homicide, manslaughter and civil liability for wrongful death and medical malpractice."[38] Although the "good medical practice" proviso may allow nontreatment of handicapped at-risk infants in some cases, it is clear that this statute demonstrates a very strong legislative intent designed to protect infants from nontreatment decisions and to enforce already existing criminal and civil penalties which impact on the process of treatment decision making.[39]

A Regulatory Approach

The federal regulations on nondiscrimination in the care of defective newborns as promulgated on January 12, 1984, even though under current challenge in the courts, as observed, declare in essence that where medical care is clearly beneficial, it should always be provided to a handicapped newborn.[40] Although recognizing a presumption should always be in favor of treatment, reasonable medical judgments will be respected regarding treatment and nourishment so long as those decisions to forgo or withhold are not made on the basis of present or anticipated physical or mental impairments.[41] Thus, decisions not to commence futile treatment which would not be of medical benefit to the infant and would in fact present a risk of potential harm will be respected.[42]

Infant Care Review Committees are encouraged, although not mandated, to be structured in the 7000 health care providers receiving federal financial assistance.[43] These committees will not only be charged with developing and recommending institutional policies concerning the withholding or withdrawal of medical treatment for infants with life-threatening conditions, but providing counsel in specific cases under prevent review.[44] Adhering to various principles approved by such groups as the American Academy of Pediatrics and the National Association of Children's Hospitals, the ICRC's will conduct their operations under the premise that where medical care is clearly beneficial it should always be provided.[45] Again, *reasonable medical judgment* will be respected regarding the bases for making decisions to forgo or to withhold treatment and nourishment.[46] Presumably, the validity of the test of reasonableness will depend upon the facts of each case-situation that arises.

Informational notices of the application of the federal law, posted where nurses and other medical professionals may view them, are required to include a statement of nondiscrimination of health services (consistent with the specific provisions of Section 504 of the Rehabilitation Act of 1973) on the basis of handicap; the notice being of a size no smaller than 5 by 7 inches, and listing a 24-hour toll-free hot line telephone number at the United States Department of Health and Human Services and/or state child protective services agency where violations of the Act may be reported.[47]

Perhaps as important as the new rules is an appendix, "Guidelines Relating to Health Care for Handicapped Infants," which, while not independently establishing rules of conduct are to be recognized as "interpretive guidelines" designed to assist in interpreting the application of Section 504.[48] Considering Appendix C(a)(1)–(3) and C(a)(5)(ii), (iii), (iv), one finds a recognition that where any of the following situational standards are in operating focus or application, *no* discrimination will be acknowledged and thus no federal intervention undertaken:

1. Where treatment based upon reasonable medical judgment would be futile
2. Where treatment is too unlikely of success given the complication of a particular case or otherwise not of medical benefit to the infant, or
3. Where there is a recognition of improbable success of a modality of treatment or risk of potential harm.[49]

It is interesting to observe that as of December 1, 1983, of the 49 cases of alleged discrimination of treatment of seriously handicapped newborns in federally assisted maternity wards, "no case resulted in a finding of discriminatory withholding of medical care."[50]

JUDICIAL POSTURING: THE BRITISH

The Sunday *Times* of December 4, 1983 carried an absorbing article concerning the plight of handicapped newborns in the United States and raised the question whether such a similar condition could ever obtain in Britain.[51] Only time, of course, can provide a definitive answer, but two important cases perhaps point to the recognition of a judicial attitude or temperament regarding the matter.

A case determined by the Court of Appeal on August 7, 1981, gives the closest indication of a judicial perspective in this area.[52] The facts showed that B, a female child, was born suffering not only from Down's syndrome but an intestinal blockage as well and would require a surgical intervention in order to relieve the obstruction if she were to live more than a few days. Although the surgery provided no guarantee of long life—in fact, there was a possibility that B might die within a few months—the evidence pointed to the fact that she could have an expectancy of normal mongol life of anywhere from 20 to 30 years if the operation was successful. Her parents decided that in "the kindest . . . interests of the child,"[53] no operation should be performed. Accordingly, they advised the doctors of this decision and it was respected. The local authority thereupon made the infant a ward of the court and sought an order authorizing the operation to be performed by other surgeons. The lower court respected the parental decision and refused to order the surgery. On appeal by the local authority, the Court of Appeal reversed and held that parental wishes were secondary to the best interests of the child. The parents made a strong argument that, owing to the fact that the child would be severely handicapped both mentally and physically, no measure of the qualitative life of a mongoloid could be evaluated properly during its predicted limited life span. The Court determined that insofar as a "happy life" could be provided a mongoloid, baby B was entitled to that life.[54]

Noting that a judicial decision in a case of this nature requires the court to consider the evidentiary proofs as well as the views of the parents and their doctors, the court acknowledged that "at the end of the day it devolves on this court in this particular instance to decide whether the life of this child is so awful that in effect the child must be condemned to die, or whether the life of this child is so imponderable that it would be wrong for her to be condemned to die."[55] The court continued that "There may be cases, I know not, of severe proved damages where the future is so certain and where the life of the child is so bound to be full of pain and suffering that the court might be driven to a different conclusion."[56]

Of interest also is Regina v. Arthur, an unreported case decided at the Leicester Crown Court on November 5, 1981, but 3 months after the In re B decision. Here, a mongoloid was born on June 28, 1980, and thereupon rejected by his parents. The consultant pediatrician, Dr. Leonard Arthur, prescribed "nursing care only" (i.e., a regime which included no food) for the child and prescribed regular doses of the drug DF118 for purposes of sedation. Originally Dr. Leonard was charged with murder, but during the course of the trial the charge was reduced to attempted murder and a subsequent decree of acquittal rendered by a jury. In his summation, the judge indicated—without apparent reference to the In re B decision—that it was lawful to treat a baby with a sedating drug and offer no further care by way of food or drugs or surgery provided two criteria are met: the child is "irreversibly disabled" *and* rejected by its parents. Thus the Arthur case would appear to suggest the issue of treatment of a severely handicapped newborn child is a private matter between physician and parent and—in light of In re B—places the law in an uncharted sphere of disequilibrium.[57] While it is clear the Arthur verdict does not legitimize the use of drugs in order to accelerate death, it is unclear whether it establishes a uniform policy of nontreatment as legal and whether "holding procedures" are valid in all cases.[58]

The American Response

Baby Jane Doe was born on Long Island, New York, on October 11, 1983, with spina bifida and an abnormally small head which was welling with excess fluid. After consultation with physicians and members of the clergy, her parents refused to allow corrective surgery. If successful, the operation might have allowed the infant to live some 20 years, but in a state of retardation, constant pain, epilepsy, and paralysis below the waist.[59]

The highest court in the State, the Court of Appeals, decided that the parents' decision must be respected. It refused to enumerate the circumstances that would trigger judicial protection of an infant of this type's interest, merely observing that there may be occasions where it would be inappropriate to intervene. Rather, it noted that the Legislature had designed a statutory scheme specifically for protecting children from abuse, at the same time safeguarding familial privacy and relationships, and that this procedure would be adhered to unless the Legislature again decided to amend the process.[60]

Although refusing to deal directly with the need to establish criteria for validating decision making in cases of this nature, a key lower court decision in New York has indicated that only if there is a "reasonable chance" for the child to lead a fulfilling and useful life, parental inaction regarding needed surgical intervention will not be permitted.[61]

The very first Baby Doe case to be found and popularized by the press involved a 6-pound baby born with Down's syndrome in 1982 in Bloomington, Indiana, who lived but 6 days. His death precipitated a national rethinking of issues of infanticide, parental decision making, and power under the Common Law to exercise jurisdiction over the care of children, and perhaps the most central issue of all: whether *quality* of life standards are more significant and fundamental than principles of *sanctity* of life. In addition to being born a mongoloid, with consequent mental retardation, Baby Doe (as he was dubbed by the press) had a malformed esophagus together with multiple physical problems. The esophageal condition prevented food from reaching the stomach. Rather than authorize corrective surgery, the parents chose to direct a withholding of food and medical treatment, save pain killers, from their son. Before an emergency appeal to the United States Supreme Court could be taken of an unwritten decision of the Indiana Supreme Court not to overturn two Monroe County Circuit Court orders preventing interference with the parental decision, Baby Doe succumbed.[62]

In an officially reported 1981 case, In re Jeff and Scott Mueller,[63] Siamese twins were born connected at the waist. The parents and their attending physicians decided against corrective surgery and efforts to feed the infants. The Illinois Department of Children and Family Services thereupon filed a petition of neglect against the parents in the Illinois Family Court and sought to gain custody of the twins. The court in due course awarded custody of the infants to the Department for the express purpose of authorizing the necessary surgery and additional medical treatment. While the court found neglect on the part of the parents in that they failed to provide treatment for the infants, it refused to impose civil or criminal liability for the neglect, which it implied occurred in an unintentional manner.[64] Subsequent efforts by the State Attorney General to secure an indictment from a grand jury against the parents and the attending physician on various charges of attempted murder, conspiracy, and solicitation to murder, met with failure.[65]

The English Court of Appeal precedent and the officially reported United States cases draw persuasive if not conclusive authority for their decisions by utilization—understood as such, or not—directly or indirectly—of a principle of a "substituted judgment." Thus the Court will seek to place itself in the position of the infant in extremis and determine whether, given its medical condition, it would wish to live under present or altered conditions; whether a meaningful or qualitative life could be achieved. Inherent in the effectiveness of application of such a principle is the employment of a cost-benefit analysis or balancing test. Or, stated simply, the costs (social, economic) of maintaining life are weighed against the benefits (religious, ethical, spiritual, etc.) of preserving it.

THE MEDICAL ATTITUDE

A startling study of special care nursing treatment of neonates undertaken at the Yale-New Haven Hospital was released in 1973 and showed that 14% of the 299 deaths recorded during the period of the study—18 months—were related to actions which withheld treatment.[66] The publication of this study initiated a public dialogue regarding the treatment of defective newborns or neonates and raised issues that heretofore had been raised privately by attending physicians, with or without familial consultation.[67]

In 1975 questionnaires were sent to all members of the Surgical Section of the American Association of Pediatricians and to all the chairmen of teaching departments of pediatrics in the United States, as well as to chiefs of divisions of neonatology and to chiefs of divisions of genetics in the departments of pediatrics. Two hundred sixty-seven physicians from the first two groups and 197 from the latter groups returned completed questionnaires. The results showed "broad support" for the propositions that: physicians need not attempt to maintain the life of every severely impaired newborn simply because the technology and the skill existed; parents and physicians (in that order) bear the ultimate responsibility for making decisions regarding the withholding or administration of treatment for handicapped at-risk newborns; such decisions should be made on the basis of the best medical predictions regarding longevity and quality of life; under certain "egregious" circumstances physicians could seek judicial intervention in order to effect treatment and, finally, decisions to treat or not to treat defective newborns were best made on a "case-by-case" or situational basis.[68]

The majority of the members of the medical profession are of the opinion that the autonomy of the parent-physician relationship should be maintained in this critical area of concern.[69] It is submitted that the affected or involved physicians and the families which they attend are the most informed parties in a case involving treatment or nontreatment decisions of handicapped at-risk newborns and that they should be accorded both respect and latitude in making these necessary decisions.[70] Contrariwise, there are a minority of physicians who maintain that parents, traumatized emotionally by the birth of a defective child, are in no position to make life or death treatment decisions regarding it.[71] No less than the Surgeon General of the United States, C. Everett Koop, has asserted that decisions that withhold treatment for handicapped newborns are acts of "infanticide."[72] He opposes the exclusive reliance that is placed upon the precincts of the physician-parent autonomy in this area of concern.[73]

The President's Commission for the Study of Ethical Problems in Medicine and Biomedical and Behavioral Research in its 1983 Report, *Deciding to Forgo Life-Sustaining Treatment*, examined critically governmental intrusions into the parent-physician decision-making process and concluded that an approach should be followed that allows for and recognizes that the nontreatment of genetically defective newborns is not unethical and should be made by the concerned parents with the advice of an attending or consulting physician, without government intervention.[74] The Commission urged that the entire process of decision making be opened to include the formation of ethic review committees and that their deliberations be considered, in order to assure an objective assessment, in those cases where the most complexity and difficulty exist.[75] Thus by the establishment of an internal review process, judicial intervention would be obviated and only be permissible when a "rapidly deteriorating medical status" of a handicapped newborn required parents and physicians alike to act without this internal review.[76]

The American Medical Association's Judicial Council reached a similar conclusion to that of The President's Commission. More specifically, the Council put forth the proposition that as to quality-of-life decisions affecting the treatment of seriously deformed infants,

> . . . the primary consideration should be what is best for the individual patient and not the avoidance of a burden to the family or to society. Quality of life is a factor to be considered in determining what is best for the individual. Life should be cherished despite disabilities and handicaps, except when prolongation would be *inhumane* and *unconscionable*. Under these circumstances,

withholding or removing life supporting means is ethical
provided that the normal care given an individual who
is ill is not discontinued. *In desperate situations involving*
newborns, the advice and judgment of the physicians should
be avialable, but the decision whether to exert maximal efforts
to maintain life should be the choice of the parents. The parents
should be told the options, expected benefits, risks and
limits of any proposed care: how the potential for human
relationships is affected by the infant's condition; and
relevant information and answers to their questions.

The presumption is that the love which parents usually have for their children will be dominant in the decisions which they make in determining what is in the
best interest of their children. It is to be expected that
parents will act unselfishly, particularly where life itself
is at stake. Unless there is convincing evidence to the
contrary, parental authority should be respected.[77] (emphasis added).

Interestingly, the Law Reform Commission of Canada reported in 1983 that decisions to treat or not to treat
defective newborns should be made according to the
medical facts of each case, be reasonable, in the best
interests of the patient, and in conformity with pertinent
standards set forth by the criminal law. Acceptable quality of life is essentially a question of fact which differs
in each case. Yet it is also a question of sound medical
judgment based in turn upon medical experience as well
as consultation with the concerned party or parties such
as parents, spouse, family, and next of kin.[78]

TRIAGE

Miracles of modern medicine may be, when performed, but curses to the recipients. A vivid description
of such a "victim" of modern medicine brings poign-

antly to the fore the major point and illustrates it graphically:

> The child lies motionless inside the plexiglass incubator. . . . She weighs 24 ounces and is two months old. Tubes, five in all, carry nutrients into her body and carry wastes away. She is covered with scabs. Her skin is yellow and slowly dying. Two weeks ago, infection ravaged her small intestine. Her body is stiff: poor circulation and blood teeming with bacteria have caused a condition similar to rigor mortis. When this little girl begins to die, she will not be resuscitated, her parents and her doctor have decided. She is taking up a bed that could be used for a potential survivor.[79]

The classical definition of *triage* may be acknowledged as being:

> The medical screening of patients to determine their priority for treatment; the separation of a large number of casualties, in military or civilian disaster medical care, into three groups: those who cannot be expected to survive even with treatment, those who recover without treatment, and the priority groups of those who need treatment in order to survive.[80]

Even before "triage" found significant application in military or civilian catastrophes, its root meaning in French—"sorting, picking, grading or selecting according to quality"—was subsequently first applied in the English language to the process of separating wool according to quality and, even later, to the separation of coffee beans into three categories: "best quality," "middling," and "triage coffee," with the last consisting of beans that had been broken and were thus the lowest in grade.[81] Over the course of time, the use of triage has been expanded to other situations where it has become, in actuality, a metaphor for social, economic, and even political decisions.[82]

Both the idea and the process of sorting casualties

of war was developed by Napoleon's chief medical officer, Baron Dominique Larrey.[83] One of the Baron's early goals in his effort to organize an efficient system of medical services to the injured was to perform surgeries as soon as possible after soldiers sustained their injuries. To this end, he developed "ambulances" whose purpose was not only to transport the wounded from the battle area but also to serve as mobile units for providing instantaneous medical assistance.[84] Additionally, he put into operation a scheme for sorting casualties on the basis of their medical need:

> Those who are dangerously wounded must be tended first, *entirely without regard to rank or distinction.* Those less severely injured just wait until the gravely wounded have been operated on and dressed.[85]

Medical personnel then were concerned centrally with finding ways to conserve scarce resources—the first and foremost being their time and their energy.[86]

Although during the Civil War the United States did not essentially classify wounded soldiers for purposes of medical treatment, but rather provided such care with regard to physical condition, during World War I it did in fact adopt from the French and the British the principle of triage.[87] And, to this day, the current military policy of the armed forces of the United States is recognized as a policy of triage which involves both the evaluation and the classification of casualties not only for purposes of treatment but of evacuation which is tied to the principle of "accomplishing the greatest good for the greatest number of wounded and injured men."[88] Thus, it is, then, that an explicit utilitarian rationale is embraced and extolled.

Principles of Allocation: Utilitarian versus Egalitarian

Since the law provides at present no uniformly agreed-upon principles that may be applied in order to regulate the allocation of scarce medical resource, current medical practice draws upon a structure for decision making evolved as such from a number of philosophical and ethical constructs.[89] There are five utilitarian principles of application operative in the hierarchy of triage: the principles of medical success; immediate usefulness; conservation; parental role; and general social value.[90] Translated as such into decisional operatives, there emerges a recognition that priority of selection for use of a scarce medical resource should be accorded to those for whom treatment has the highest probability of medical success, would be most useful under the immediate circumstances, to those candidates for use who require proportionally smaller amounts of the particular resource, those having the largest responsibilities to dependents, or those believed to have the greatest actual or potential general social worth.[91] The utilitarian goal is, simply stated, to achieve the highest possible amount of some good or resource.[92] Thus, utilitarian principles are also commonly referred to as "good maximizing strategies."[93]

Egalitarian alternatives, contrariwise, seek either a basic maintenance or a restoration of equality for persons in need of a particular scarce resource.[94] There are five basic principles utilized here: (1) the principle of saving no one—thus priority is given no one because, simply, none should be saved if not all can be saved; (2) the principle of medical neediness under which priority is accorded those determined to be the medically neediest; (3) the principle of general neediness which allows priority to be given to the most helpless or generally neediest; (4) the principle of queuing where priority is given to those individuals who arrive first; and (5) the principles of random selection where priority of selection is given to those selected by pure chance.[95]

To the utilitarian, maximizing utility, and hence what is diffusely referred to as the "general welfare," are both the primary ground and subject of all judgments.[96] That which is required in order to maximize utility overall may thus infringe upon an individual's own entitlements or rights to particular goods.[97] Accordingly, moral rights are either rejected generally or recognized as certainly not absolute.[98]

Philosophy and religion may well provide us all with the necessary balance and direction for life and allow us to develop an ethic for daily living and a faith as to the future; but in cases of neonatology where law, science, medicine, and religion interact, great care must be exercised in order to prevent inexplicable fears and emotions—often fanned by journalistic prophets of the "'what if" shock culture—taking hold of and thereby blocking powers of rationality and humanness.[99] The basic challenge of modern medicine should be, simply, to seek, promote, and maintain a level of real and—when the case may dictate—potential achievement for its user-patients which allows for full and purposeful living.[100] Indeed, man himself should seek to pursue decision-making responsibilities and exercise autonomy in a rational manner and guided by a spirit of humanism. Further, he should seek to minimize human suffering and maximize the social good. Defining the extent and application of the social good will obviously vary with the situation of each case.[101]

Balancing Costs and Benefits

The conundrum of seeking to maintain purposeful living yet at the same time protect the recognition of the sanctity of life finds reality and force when dealing with the plight of genetically defective newborns. This conundrum is also to be recognized as presenting a duality of goals. One goal is and must be balanced against another in attempting to reach a level of distributive justice in the hard decision required here. The situation ethic must be predominant over a harsh, unyielding a priori standard. Viewed from another perspective, this balancing test underscores recognition of the fact that human life is in actuality but a resource, as are natural, physical, and environmental resources. Thus the primary goal for the conservation of every resource is the maximization of its full use or potential, be it viewed as economic, social, cultural, or political. Waste must be avoided. Considered as such, then, in seeking to maximize the good of this precious resource of life, the right of personal autonomy and spiritual awareness are but vectors of forces that must be additionally factored into any balancing equation. State interest is yet another positive force and also a constraint on autonomous or, in this area, parental-familial medical decision making.[102]

When considering the severely defective newborn, the costs to the individual if maintained vis-à-vis the quality of life and extent of life must be weighed against not only the side effects on the parents at a social, emotional, and economic level, but on the hospital staff in the nursery watching the infant die and listening to its strangled cries[103] as well as on society as to the loss of a young citizen and the potentially dangerous recognition that death may have for future similarly disposed citizens. Indeed, the goal of achieving a manageable level of sustenance may well involve incalculable levels of suffering for both the active and passive participants.[104]

Lord Justice Ormrod, a qualified medical practitioner and a Fellow of the Royal College of Physicians, observed in 1978 that the "cost-benefit" equation was no longer ignored in making modern medical treatment decisions.

> The medical profession has . . . recognized that it is concerned with something more than the maintenance of life in the sense of cellular chemistry, and so implicitly accepted the concept of 'quality of life' from which it has in the past always fought shy, for obvious reasons. It has implicitly accepted that considerations of cost-benefit cannot be completely ignored. . . . Ten or fifteen years ago, mere mention of either was enough to precipitate an emotional response from non doctors. Now they are explicit and can be discussed and debated rationally—an important advance from many points of view.[105]

THE ROLE OF ETHICS COMMITTEES

Ethics committees have been utilized, with one or another degree of success as either advisory or enforcing, since the 1970s.[106] In fact, a 1976 decision by the New Jersey Supreme Court was the progenitor of these committees.[107] There, it was determined that ethics committees, composed variously of physicians, social workers, attorneys, and theologians, endeavor to review the circumstances of specific ethical dilemmas in hospital settings normally involving the issue of whether to maintain or remove life-sustaining treatment modalities and thus assist both patients, their families, and the attending physicians in finalizing decisions and, at the same time, conditioning civil or criminal liability that might otherwise arise from such undertakings.[108]

Sparked by the Infant Doe case in Indiana[109] and the Baby Jane Doe case in New York[110] involving the birth of genetically handicapped newborns, together with federal administrative regulations promulgated by the United States Department of Health and Human Services,[111] a new and sustained level of interest has emerged in both refining and utilizing a variant of ethics committees.

While the federal regulations apply only to handicapped at-risk newborns, there are strong indications that ethics committees are being formed to assist in aiding situations other than critical ones found in neonatal units.[112] California, New York, New Jersey, Connecticut, and Massachusetts are states whose hospital systems have responded in advancing the development of ethics committees that participate in, one level or another, the clinical decision-making process.[113]

Even though admittedly becoming more popular, even the proponents of ethics committees recognize clearly that the committees should *not* become an ultimate decision-making body.[114] Rather, they would "strive to emphasize the role of attending physicians and surrogates as the primary decision makers for incapacitated patients."[115] Yet, depending upon the nature of the particular case under review, the degree of patient incompetency, level of informed or proxy consent given, and rationality of reasons given for refusing medical treatment, there should be an understanding that the ethics review committee might in fact require a level of authority to order a (temporary) postponement of decisions which it has counseled against or even to initiate judicial proceedings in order to seek a review of decisions which it opposes.[116]

The positive values that ethics committees provide can be found not only in their structuring of a sound system for careful, human decision making, but in their services as a valuable bridge "between societal values and the actual developments occurring in the institutions that care for and treat the particular patients who manifest these dilemmas."[117] Not only do they assist in efforts to distinguish between ethical dilemmas where consensus may exist and in those cases where none is achievable, but also in developing policies and guidelines designed to facilitate decision making, advancing an educative component targeted at the medical and nursing staffs which would better enable them to both comprehend and to resolve ethical dilemmas arising from the administration of modern health care.[118]

Finally, the institutional ethics committees would serve as a consultative and case review mechanism whereby its individual members would be available to discuss the ethical and social concerns of interested parties, and by having the committee provide advice to parties that seek it.[119] The scope of individual case review would, as noted previously, necessarily vary with the facts of each case under consideration. Some cases would require a more aggressive posture on the part of the committee than others.

The value of such committees will increase if courtroom adversarial situations can be prevented. Toward the achievement of this end, government regulation in this field should be avoided. In its place, concerted efforts should be made toward following the example of hospice care in the United States, which developed as local and independent, but connected entities in the health care system, rather than follow the path of institutional review boards, which were created by federal regulation to protect the subjects of human biomedical experimentation.[120] If concerned individuals in present health care institutions, be they pediatric, acute care, or geriatric, work toward establishing a viable structure designed to protect both patients' rights and ensure that reasonable and fair decisions are guaranteed for those unable to decide for themselves, the advances of governmental regulation will be halted and the goal of improving patient care will be enhanced.[121]

SEEKING A CONSTRUCT FOR DECISION MAKING

The underlying principle of application should always be to minimize suffering and maximize the qualitative potential for fulfilling human relationships, thereby promoting a purposeful life for the at-risk infant.[122] The extent to which this principle, inquiry, or test should be applied depends solely upon the facts of each situation as it arises to present a problem. To have an unyielding a priori standard of mandated care for all seriously handicapped newborns would be unjust not only for the infant itself, and promotive of its undue suffering, but equally unjust and harsh for its parents; and would present an unreasonably heavy economic burden to society for its maintenance and allocation of scarce and expensive medical support resources and mechanisms. Efforts must always be made to ensure, however, that if a class is structured and labeled, "disabled," it is drawn as *narrowly* as possible and is as strictly *defined* as possible.[123] Thus the overriding issue is whether a construct can in fact be so designed as to assist the supervising physicians, the family and their religious counselors, and the state (when involved) in defining the parameters of a class of nonsalvageable defective newborns.

At various times it has been suggested that the lack of capacity for consciousness,[124] social interaction, human relationships (especially love),[125] and rational thought were the four most important considerations in determining who was to be placed in the "nonsalvageable" classification.[126] The importance of each capacity in the hierarchy of the classification depends, very obviously, upon one's particular social, ethical, religious, and philosophical perspective. One leading ethicist has stated that ". . . the warmth of human interaction, the love of one person for another, the emotional bonding that links people in moral communities does not require a capacity for consciousness."[127]

What is crucial in assessing these various capacities, no matter which are regarded as more central or primary by the philosopher-ethicists, is the actual physical condition of the handicapped newborn. If some agreement or at least a consensus could be reached that certain genetic afflictions were not significantly correctable by surgery or medical treatment so as to promote a life free of intense pain and suffering, then better, more informed decision making could be considered by the family and its expanded circle, aided by the medical recognition or determination that the at-risk infant was one member of a classification for whom sustained living would be inhumane.

The Sheffield Criteria

The closest possible effort to developing a classification or construct may be found in the results of a spina bifida study undertaken at Children's Hospital in Sheffield, England, in the early 1970s where a list of six defects were found and agreed upon as precluding the possibility of an independent, dignified life—or, for that matter—of the enjoyment of meaningful interpersonal relations. They are:

1. Thoracolumbar or thoracolumbosacral lesion.
2. Gross paralysis with a neurologic segmental level at L3.
3. Kyphosis or scoliosis.
4. Gross hydrocephalus with a head circumference at least 2 cm above the 90th percentile related to birth weight.
5. Other gross congenital defects, such as cyanotic heart disease.
6. Intracranial birth injury.[128]

The New South Wales Guidelines

The Department of Health of New South Wales, Australia, has drafted admission guidelines for neonatal intensive care units which may be considered not only humane, but economically efficient. They state that:

1. The triage principle should be adopted in the neonatal intensive care field.
2. Newborn babies weighing more than 750 grams should generally be accepted for treatment unless they are suffering from serious malformations (including some chromosomal abnormalities) that will seriously impair their quality of life.
3. Newborn babies who, despite skilled resuscitation and exclusion of reversible problems, have not attained their own cardiac output after 10

minutes have elapsed should, for humanitarian reasons, be carefully considered for triage.

4. Some babies weighing more than 750 grams may subsequently be excluded because of severe intraventricular hemorrhage or severe brain damage.

5. Infants with severe abnormalities of cardiac and/ or respiratory systems (hypoplastic left heart syndrome, laryngeal and tracheal aplasia) where surgical correction is not feasible should be considered for triage.[129]

An Expanded Criteria

It has been suggested that termination of a pregnancy during the third trimester (in the twenty-fifth week) can be morally justifiable—that is, permissible in accordance with the mother's wishes—when two conditions are fulfilled. The conditions are that: (1) the fetus be afflicted with a condition that is (a) incompatible with postnatal survival for more than a few weeks, or (b) characterized by the total or virtual absence of a cognitive function, and (2) that highly reliable diagnostic procedures be available for determining prenatally that the fetus in fact fulfills either conditions 1(a) or 1(b).[130] One entity, anencephaly, or the markedly defective development of the brain, together with the absence of the bones of the cranial vault, clearly fulfills both conditions.[131]

Several other disorders fulfill condition 1(a)—that is, a fetus thus afflicted cannot survive for more than a few weeks after birth—renal agenesis, or failure or imperfect development of the renal system, infantile polycystic kidneys with resultant hypoplastic or under developed lungs, and Meckel's syndrome.[132] The difficulty here is that these disorders do not fulfill the *second* condition, in that there are no highly reliable diagnostic procedures available to determine prenatally the presence of the disorder.[133]

On the other hand, Tay-Sachs, a genetic anomaly, can be reliably diagnosed in utero, but an afflicted infant may well and usually does in fact have a number of months of normal life before rapid deterioration occurs and thus does not fulfill the first condition, namely that postnatal survival be for no more than a few weeks or be characterized by the total or virtual absence of cognitive functions.[134]

Low Birth Weights

A report of infant births in the United States disclosed that infants in the 501 to 750 gram range (1 lb., 1½ oz. to 1 lb., 10¼ oz.) are often treated aggressively; those in the 751- to 1000-gram range (1 lb., 10¼ oz. to 2 lb., 3 oz.) are also commonly treated in an aggressive manner, while those infants weighing more than 1001 grams (2 lb., 3 oz.) at birth are *routinely* treated aggressively.[135]

Unless substantive reasons exist for the withholding of treatment, intensive care in Britain and Sweden is generally reserved for infants over 750 grams. Similarly, when an infant weighs less than 750 grams, it is seldom subjected to aggressive care (e.g., machine-assisted respiration). As a consequence of this practice or unstated policy, fewer disabilities result in British and Swedish infants of low birth weight from aggressive treatment in the United States. Interestingly, the report concluded that the major difference between infants provided with intensive care in Britain and Sweden and

the United States was not based on genetic grounds—
for which there is considerable consensus—but, rather,
on the weight below which aggressive therapies could
be withheld.[136]

Since the infant with a very low birth weight is susceptible not only to brain injuries which in turn may
result in associated handicaps such as mental retardation and cerebral palsy, and his sustained outcome can
only be obtained at considerable financial expense, it has
been suggested that a cut-off weight of 1000 grams be
set (or, in other words, about 2 lb., 3 oz.) below which
aggressive treatment could be withheld justifiably. Another suggestion would be to withhold aggressive infant
care when a birth weight of 1000 grams or less is
recorded for those born in a state of severe asphyxia-
tion.[137]

It should be noted with clarity that nonaggressive
care does *not* mean *no* care. Rather, it has been termed
conservative care and is recognized as a less intensive
modality or therapy designed to promote comfort and
well-being by, for example, keeping a distressed infant
warm, providing fluids for him when indicated and,
where necessary, placing him under an oxygen hood in
order to prevent cyanosis.[138]

Underlying these suggestions is the ultimate realization that selective nontreatment would be administered on the basis of a determination made by the family
and its physicians regarding the potential for quality of
life if sustained and developed and of economic consideration of cost-effectiveness in saving and promoting a
particular life. Thus an infant who, from a neurophy-
siological consideration, has no conceivable potential to
engage in rational deliberations and from a sociode-
mographic index would be unable to enjoy a qualitative

life would not be assisted in an aggressive manner with its struggle for life.[139]

Perhaps the most commonly agreed-upon genetic factors that would justify the withholding or the withdrawing of aggressive therapies are tied to severe abnormalities, diseases, or damages to an infant's central nervous system, especially the brain. More specifically, such agreement would include cases of hydroanencephaly, severe neural tube defects, gross hydrocephalus (if complicated by infection) and specific chromosomal disorders such as trisomy 13 and 18. Additional cases might include infants with extensive and fully documented brain damage after (asphyxia and) hemorrhage.[140]

CONCLUSIONS

Life, viewed as a human resource, should be developed and preserved along those lines that allow for the achievement of its fullest potential for total economic realization, maximization, or productivity. Indeed, human life, at whatever state of development, is both a precious and sacred *resource*.[141] Its initial advancement or abrupt curtailment should be guided always by a spirit of humanism. Viewed thus, attainment of the quality of purposeful, humane living becomes a coordinate or complement to total economic utility.[142]

Child protection laws are, of course, necessary. Their design and promulgation by the government at the state and federal levels are crucial if standards of equal protection for all of its citizens—regardless of age or physical stature—are to be assured. It is a dangerously thin line to tread between familial privacy in decisuion-making matters of this nature and government intervention.[143] The judiciary, when called upon to evaluate cases of alleged abuse of handicapped newborns, can be aided by a close working partnership with the medical profession in seeking to determine those situations where the withholding of needed medical or surgical modalities of treatment would be in the best interests of the infant, as well as all others immediately concerned.[144] This is a proper judicial inquiry and a proper role for it to pursue.

The construct for decision making that I have proposed here is of value not only to the courts, but also to the parents of a handicapped at-risk newborn who themselves must confront the initial decision regarding the absolute sanctity of all life or a standard of qualitative living; a decision made either alone, with attending physicians or, where permitted, with the guidance of an ethics review committee. Aided then by medically agreed-upon evaluations of genetic and other birth deficiencies, as for example the Sheffield List, the New South Wales Health Department regulations on Neonatal Intensive Care Unit Admissions, and birth weights for neonates that preclude qualitative, purposeful living, together with the operation in hospitals of ethics review committees, decisions can and should be made within the bounds of familial privacy and the confines of the doctor–patient relationship of confidentiality, all free as such from governmental interference.

EPILOGUE

On April 15, 1985, a new final rule was issued by the United States Department of Health and Human Services regarding the care of handicapped at-risk newborns.[145] The meaning of the term "medical neglect" here includes "the withholding of medically indicated treatment from a disabled infant with a life-threatening condition,"[146] and designates the child protective service in each state as the proper body to enforce an administrative policy that demands no infant be the victim of "medical neglect."[147]

The precise definition of what it means to withhold medically indicated treatment is:

> the failure to respond to the infant's life-threatening conditions by providing treatment (including appropriate nutrition, hydration, and medication) which, in the treating physician's (or physicians') reasonable medical judgment, will be most likely to be effective in ameliorating or correcting all such conditions. . . .[148]

The three exceptions where withholding of treatment "other than appropriate nutrition, hydration or medication" are *not* "medical neglect" are set forth as when:

> (i) The infant is chronically and irreversibly comatose;
> (ii) The provision of such treatment would merely prolong dying, not be effective in ameliorating or correcting all of the infant's life-threatening conditions, or otherwise be futile in terms of the survival of the infant; or
> (iii) The provision of such treatment would be virtually futile in terms of the survival of the infant and the treatment itself under such circumstances would be inhumane.[149]

This new rule has been criticized as merely symbolic and but a "punctuation mark rather than a new page in the story of Baby Doe."[150] While the rule supports the trend toward aggressive treatment for life-threatening medical problems in infants having additional nonlethal physical or mental disabilities (e.g., infants such as the original Baby Doe), and articulates the "best interest of the infant" standard in making treatment decisions, and "encourages" hospitals receiving federal funding to establish Infant Care Review Committees, there remain some serious problems.[151] For example, this rule, despite rather liberal interpretive guidelines,[152] maintains an ambiguity inherent in the enacting legislation itself. The rule states that treatments falling within the description of "inhumane" are not mandatory.[153] The average physician is expected to exercise "reasonable medical judgment" when choosing to offer a modality of treatment that might well be "virtually futile" when the at-risk infant's hope of survival is doubtful. "Virtually futile" is not taken as meaning absolute certainty, the prospect of death not "imminent," but to "threaten the life of the infant in the near future" instead of "the more distant future."[154]

Another area of concern where treatment could well be affected is in the withholding of either nutrition or fluids. Although its listing of exceptions to the requirement to treat infants is stated, the rule provides specifically that "appropriate nutrition, hydration and medication" are to be given if the infant is either comatose,[155] dying, or if other treatment is virtually futile.[156] This could very well leave open the distinct possibility of withholding nutrition, hydration, or medication on the very grounds that they are not "appropriate" under the circumstances.[157]

Only the course of time will determine the extent to which this new rule meets a critical societal concern or shows itself to be but a shallow and symbolic palliative. The problem of the handicapped at-risk newborn will not go away. It can be attacked by the legislatures or the courts. The hope is that judicial restraint and deference will prevail and acknowledge the legislative forum as the better one for dealing with this grave social issue.

A Determinative Judicial Resolution?

On June 9, 1986, Mr. Justice John Paul Stevens, writing for the United States Supreme Court, held that federal regulations designed to prevent hospital discrimination in the care of handicapped newborns and promulgated under Section 504 of the 1973 Federal Rehabilitation Act were invalid and thus not authorized by the Act itself.[158] The Court accepted the argument that I made in my *Amicus Curiae* Brief in this case that causes of action designed to prevent the discrimination of defective newborns were traditionally relegated to state law in an area that is basically the concern of the state—namely, the prevention of child abuse. Indeed, each of the 50 states and the District of Columbia has enacted legislation designed specifically to provide for a duty to report and a remedy to investigate, take custody, and prosecute when necessary incidents of "child abuse" occur.[159] The Section 504 Regulations were thus held by the High Court to be duplicative, unnecessary, redundant, and intrusive into the inherent rights of the states to regulate this area.

The pertinent part of the Federal Child Abuse Amendments of 1984 as they relate to defective newborns structure a framework for medical care and treatment.[160] Under these amendments, beneficial medical treatment must be provided by hospitals receiving federal grant assistance to any infant unless it is chronically and irreversibly comatose; the provision for such treatment would merely prolong dying or be otherwise futile in terms of survival, and the treatment provision would be futile and the treatment itself would, under the particular facts of the case in question, be considered inhumane.[161] The codified amendments state specifically that even under these exceptions, a patient cannot be denied "appropriate nutrition, hydration and medication."[162] It has been suggested that this standard provides a workable foundation for the development of a similar standard of care for terminally ill adults.[163]

Chapter 4

The Contemporary Influence of Genetics and Eugenics in Family Planning

INTRODUCTION

Substantial scientific evidence exists which indicates man's genetic inheritance acts as a major influence not only upon his behavior but on his health.[1] In the United States, for example, it is estimated that one out of every 20 babies is born with a discernible genetic deficiency;[2] of all chronic diseases, between 20 and 25% are predominantly genetic in origin.[3] At least half of the hospital beds in America are occupied by patients whose incapacities are known to be of a genetic origin.[4] Since modern medicine can alleviate the symptoms of some genetic disease syndromes through sophisticated treatment, many who are afflicted with genetic disease and who in the past would not have survived, are now maintained for extended periods. Medicine is unable to do much by way of *curing* genetic defects,[5] however, and those afflicted with genetic diseases who are kept alive by modern technologies can reproduce and thus may increase the number of defective genes in the genetic profile of the human population.[6]

Considerable research into techniques for perfecting genetic engineering has been undertaken in an attempt to develop new, effective treatment for individuals with inherited diseases.[7] Under the rubric of the New Biology, scientists are both investigating and developing many interventions, including gene deletion surgery, splicing and transplantation, cloning, in vitro or test tube fertilization, embryo implantation, parthenogenesis, amniocentesis, and experimentation with the scope and application of DNA.[8] Genetic engineering utilizes some of these procedures to reorganize human genes to produce varied, particular characteristics.[9]

In order to combat genetic disease, genetic engineering may rely—and in fact frequently does—upon eugenics, the science that deals with the improvement of heredity. Stated simply, a positive eugenics program seeks to develop superior qualities in man through the propagation of his superior genes;[10] with the positive eugenicists seeking to produce a "new breed" with keener and more creative intelligence.[11] On the other hand, a negative eugenics program attempts only to eliminate genetic weaknesses.[12] When seen in application, positive eugenics programs encourage the fit and "proper" individuals to reproduce, while negative eugenics programs discourage the less fit and those with inheritable diseases from procreating.[13] Abortion is one way of implementing a program of negative eugenics after earlier measures of regulation have failed.[14]

My purpose in this essay will be to explore the extent to which positive and negative relationships or compatibilities and incompatibilities exist and are found within eugenics as a directive force in the science of genetics and to thereby test the extent of their dependence or their independence as an influence in modern family planning.[15]

THE HISTORICAL PERSPECTIVE

Plato, in his *Republic*, idealized selective breeding as the foundation for the creation and maintenance of a superior Guardian class.[16] After postulating a theory of evolution which was based upon the natural selection of the fittest organisms by virtue of their greater reproductive successes in the competitive struggle for existence in 1859 with his treatise, *On the Origin of Species*, Charles Darwin went on to suggest 12 years later in *Descent of Man and Selection in Relation to Sex* that man could profit if selective breeding techniques were introduced into his reproductive cycle.[17] It remained for his cousin, Sir Francis Galton, however, to achieve the status and recognition of being the true father of eugenics in 1883.[18] As early as 1869 Galton had asserted that each generation had a power and a coordinate responsibility to those that followed to use their natural gifts so that they would be of measured advantage to future generations.[19] As formulated as a theory in 1883, "eugenics" was denominated as a scientific approach to give "the more suitable races or strains of blood a better chance of prevailing speedily over the less suitable than they otherwise would have had."[20]

First in Europe, and subsequently in the United States, social reformers and modernists seized upon Darwin's theory of evolution as a key to understanding the social disorganization of the period.[21] Indeed, Darwinists were formed as a group that saw the decaying social order as the product of a type of healthy competition where only the fittest survived.[22]

The real honor of being the "father" of modern genetics fell to Gregor Mendel, an Austrian monk who in the late 1860s began exhaustive experiments into inheritance factors that were later designated as genes or units of heredity.[23] Mendel discovered, through a process of crossbreeding peas, that a pair of determiners or genes were the mechanisms through which inherited traits were passed. Thus if a plant were to inherit a gene for round leaves from each parent, it would have that specific trait. Yet, interestingly, where a plant might inherit one gene for sets of round leaves and another gene for pointed leaves, in this case the plant would exhibit but one of those traits; and the gene for that trait would be considered the dominant gene, the classified as recessive. Recessive traits would only appear when a plant inherited two recessive genes. Accordingly, a recessive trait could "skip" a generation yet be expected to appear subsequently in another. Using this data, Mendel went on to develop a detailed system of ratios which was used to describe the appearance of a trait.[24]

While Mendel sought application and validation of his ratios only as to peas, the eugenicists proceeded to blanket use of these ratios in order to describe evolutionary genetics at a time scientific knowledge was quite primitive. Almost all of an individual's physical and psychological characteristics were attributed to the presence in his parents' reproductive, or germ cells, of a gene for each specific trait. While little disputation was entertained as to the inheritability of such common physical traits as iris and hair pigment, or skin color, the eugenicists extended their position by maintaining that psychological traits such as sincerity or insincerity and truthfulness or untruthfulness were also inherited.[25]

While the noble ideals of positive eugenic programs sought to encourage those endowed with what were perceived as socially beneficial traits to take basic eugenic principles into consideration when choosing a marriage partner, as well as family size, the negative program for eugenic improvement stressed eradicating socially inadequate germ-plasm (e.g., the feebleminded) from the American stock through legally sanctioned sterilization procedures.[26] This program captured the interest and the imagination of a large number of Americans, while the nobility of purpose and idealism seen in implementing a positive eugenic program never really developed or, for that matter, flourished.[27]

In 1929, those determined to be "socially inadequate" and recognized as the target groups for sterilizations were: the feebleminded; the insane (which included the psychopathic); epileptics; inebriates (which included drug addicts); the diseased (e.g., the tubercular, syphilitic, leprous, and all others with chronic infectious and legally quanrantined diseases); the blind and those with seriously impaired vision; the deaf and those with seriously impaired hearing; the deformed (which included the crippled); and dependents such as orphans, ne'er-do-wells, the homeless, tramps, and paupers.[28] The stated goal of a number of the eugenicists was to build sufficient institutions so that by the year 1980 care could be "extended" to the 1500 feebleminded per 100,000 of the population that, it was maintained, would then be living in the United States.[29]

By 1925, 25 states had enacted at least one piece of eugenic sterilization legislation. While varying classes of people were declared to be subject to the laws, each law combined various degrees of punitive, eugenic, and therapeutic motives to effectuate its intent.[30] Various court challenges to the constitutionality of the statutes were maintained and when such a statute of this type was in fact determined to be unconstitutional it was a decision founded on a denial of equal protection of the laws (i.e., an invidious discrimination of an existing class of citizens), a violation of the due process of laws guarantee of the Constitution, or a recognition that the sterilizations were cruel and unusual punishment.[31]

Although by 1931 some 32 states had passed one or another type of sterilization legislation, the full popularity of the eugenics movement had begun to decline as early as 1927.[32] Interestingly, during the 1920s new scientific investigations began to show clearly that feeblemindedness was not a direct consequence of Mendelian ratios, but rather the result of very complex etiologies.[33] Finally, then, in the full decade of the 1930s in America, not only did more startling research advance in psychology, sociology, and anthropology show with clarity that environmental surroundings were certainly *as* significant a determiner of human character and intelligence as heredity, but, as important, the passionate commitment of the original leaders of the eugenics movement was not found to be replaceable in the new converts, once the original leadership ranks were thinned by death or retirement.[34]

IMPLEMENTING A NEGATIVE EUGENICS PROGRAM

In seeking to eliminate genetic weaknesses from society, a negative eugenics program necessarily requires some process to determine genetic composition. Genetic screening and counseling accomplish this objective by identifying carriers of genetic diseases and advising couples whether reproduction is biologically desirable.[35] Such screening and counseling may occur at both preconceptual and postconceptual stages.[36] A simple preconceptual screening procedure consists of withdrawing and analyzing a blood sample in order to determine if an individual possesses any recessive traits for a genetic disease.[37] Postconceptual screening and counseling procedures are more medically complicated and also pose more complex legal issues.

Amniocentesis

A recently developed postconceptual screening procedure, amniocentesis, has emerged as a principal element of negative eugenic programming. The procedure consists of inserting a needle through the abdominal wall of a pregnant woman into the amniotic sac containing the fetus, withdrawing a sample of the sac fluid, and analyzing it.[38] Since the sac contains cells from different parts of the fetus, analysis of this sample reveals the sex of the fetus and also whether it will be affiliated with certain genetic disabilities.[39] By permitting a physician to predict accurately the presence of certain genetic defects, amniocentesis significantly advances standard genetic counseling procedures that must rely on probabilities.[40]

If amniocentesis reveals a genetically defective fetus, the parents face the difficult choice of whether to abort it. A couple informed of a genetically defective fetus may decide for religious, personal, or ethical reasons that they want to guarantee the birth of the life they created and therefore allow the pregnancy to continue. Such a choice raises the issue whether the child could bring a tort action against his parents for wrongful life. Under current law, such a claim would likely fail.[41]

Genetic Screening and Counseling Programs

Some of those concerned with negative eugenics currently have emphasized the need for the wide application of traditional screening procedures to identify the carriers of certain diseases.[42] Certain leaders of Jewish communities, for example, encourage citizens of their communities to participate in screening to identify carriers of the Tay-Sachs recessive gene, which can cause a debilitating illness.[43] Federal legislation permits the use of public funds to establish voluntary genetic screening and counseling programs for carriers of sickle-cell anemia;[44] and some state legislatures have gone farther and required genetic screening of school-age children for the trait.[45] New York provides for premarital testing to identify carriers of the same defective gene.[46] Genetic screening programs also may include provisions for counseling.[47] Unfortunately, counseling efforts to date have been sporadic and ineffective.[48] If genetic screening programs are to have any significant impact, more effective counseling techniques must be devised and implemented.[49]

Public acceptance of mandatory genetic screening programs should not be impossible to achieve. Premarital genetic screening would be an easy addition to state statutes that presently require premarital testing for maternal rubella titre (although not itself considered to be a genetic defect), blood group, and Rh status.[50] One scholar asserts that statutes requiring genetic screening for the population at large would be a simple and readily acceptable extension of present laws requiring vaccinations and chest X rays for school children.[51] Moreover, societal problems such as population control, the cost of supporting the handicapped, and the general welfare of the population favor the trend toward mandatory genetic screening.[52]

Other legal scholars maintain that compulsory genetic screening programs may be unconstitutional.[53] They assert that the taking of a child's blood sample would constitute a physical invasion of the body in violation of the Fourth Amendment to the Constitution and that a compulsory counseling program would interfere with the fundamental rights to marry and procreate.[54] These critics also contend that a less intrusive voluntary program, together with extensive dissemination of educational material, could accomplish the same objectives.[55] Although genetic screening involves a minor intrusion into an individual's body and may involve a "search" within the meaning of the Fourth Amendment, the search is not unreasonable and prohibited if executed in a proper manner and justified by a legitimate state interest.[56] Similarly, assuming *arguendo* that mere screening and counseling interfere with the right to procreate, such interference may be justified by a compelling state interest that must be preserved. The state's interest in improving the quality of a population's genetic pool in order to minimize suffering, to reduce the number of economically dependent persons and possibly to save mankind from extinction arguably justifies the infringement of individuals' civil liberties.[57]

Unfortunately, voluntary programs have little value in achieving the purposes for which they are structured. People are too preoccupied with the daily vicissitudes of life to be concerned with prospective occurrences of genetic possibilities. Therefore, although a voluntary program is conceded less intrusive, the only way to achieve positive, enduring results is to implement some form of mandatory genetic screening program.[58]

Restrictions on Marriage

An even more effective means of preventing the birth of genetically defective persons is prohibiting marriage between carriers of the same genetic defect. Both constitutional and social objections have been raised to such a restriction on marriage.[59] Existing laws prohibiting marriage for eugenic reasons and proposals to restrict marriage between carriers of the same genetic defect are attacked as being excessively broad, and the suggestion is made that only procreation needs to be regulated to ensure both eugenic preservation and responsible parents.[60]

Since procreation traditionally is set within the marriage framework, however, establishing restrictions on marriage is the most practical mechanism for implementing a negative eugenics program. Moreover, married couples prohibited from procreation nonetheless might have children accidentally or intentionally.[61] Whether a state's pursuit of the public's health and welfare would justify an abridgement of the fundamental right of marriage between carriers of the same genetic defect is doubtful. Such restrictions also might well prove ineffective in the contemporary atmosphere that is increasingly tolerant of free love and common law (or de facto) relationships. Thus, it is unlikely that restrictions on marriage would prove to be an acceptable method of eugenic control.

Restrictions on Reproduction

Modern cases support the proposition that marital and procreative decisions fall within a constitutionally protected zone of privacy.[62] As long ago as 1941 the

United States Supreme Court declared that man possesses the basic civil right to have offspring.[63] More recently, the Court has held that the choice of whether to give birth is within a constitutionally protected zone of privacy.[64] These broad pronouncements do not force the conclusion, however, that all restrictions on reproduction are not per se unconstitutional. If a state may prevent a person from marrying more than one person at a time, should it not have the same power to prevent a person from having more than one or two children? The right to procreate may not include a right to breed without some restrictions.[65] Societal interests may be sufficiently powerful to justify at least some regulation for limitations on reproduction.[66]

Some legal precedents do uphold the constitutionality of eugenic sterilization. In the yet to be overruled Buck v. Bell,[67] the Supreme Court of the United States upheld a Virginia statute providing for sterilization of inmates committed to state-supported institutions who were found to have a hereditary form of insanity or imbecility.[68] And a number of the states have some form of compulsory sterilization legislation,[69] with the courts typically upholding the validity of actions brought thereunder.[70]

The extension of Buck to sterilization of carriers of recessive defective genes cannot be accomplished without difficulty. Since its decision in that case, the Court has increasingly recognized the right to marry and have children as a basic or fundamental right and that a state must show a compelling interest in order to justify any abridgement of the right itself.[71] Several factors seem to indicate that the state interest is not as compelling with regard to sterilization of carriers of defective genes as it is with regard to mental incompetents. A mental incompetent may well be unable to be an adequate parent, and the burden of care therefore would fall upon the state.[72] Moreover, the sterilization of mental incompetents in institutions can be said to benefit them directly in that it "enable[s] those who otherwise must be kept confined to be returned to the world . . . ."[73] The Court seemed to have assumed in Buck, however, that there is a strong likelihood that the child of an intellectually defective mother would in fact inherit the same defect;[74] even though the child of two heterozygous individuals had only a one-in-four chance of exhibiting that defective trait.[75]

The distinguishing features of Buck v. Bell do not indicate that the state cannot offer compelling justification to warrant mandatory restriction on reproduction. Such justifications can be found in society's interest in the reduction of human suffering, and in safeguarding the health and welfare of its citizens in the allocation of economic resources and in population control.[76] In Buck, Justice Holmes stressed that "it would be better for all the world . . . if society can prevent those who are manifestly unfit from continuing their kind."[77] Perhaps world conditions have become so complex and resources so valuable that society now has a compelling interest in restricting reproduction by those who, although *not* "manifestly unfit" themselves, perpetuate human suffering by giving birth to genetically defective offspring.

THE AUSTRALIAN POSTURE

Sterilization

In contrast to the United States and Canada, in Australia there are no compulsory sterilization laws directed toward restricting those from propagating who are suspected of carrying deleterious genes or diseases.[78] Indeed, if there is any kind of procreative "policy" in Australia, it is in the encouragement of reproduction, not its restriction.[79]

The availability of information concerning the frequency of sterilization of mentally retarded citizens in Australia is difficult to obtain since this surgical intervention is not usually conducted on residents of institutions, but rather on those in private residence with their families.[80] State government institutions in New South Wales reported however that in 1979, for contraceptive or hygienic reasons, two or three tubal ligations were performed; and no more than five hysterectomies had been performed during the last 20 years.[81]

Genetic Counseling and Screening

Through genetic counseling, as has been observed, prospective parents learn the likelihood of a disease they may carry genetically being passed on to one of their offspring. Most often, the critical information needed to construct family histories is to be found only in various medical records. Under present Australian law, family members have no absolute right or access to their own medical records and thus a genetic counselor's standards of "probability" may be inaccurately skewed one way or the other without the benefit of a complete family medical profile.[82] Approximately 90% of those couples participating in a program of genetic counseling have either had a handicapped child or know of one in their immediate family.[83] Tragically, the level of communication and of retentive understanding is perhaps the greatest impediment to an effective utilization of counseling here. It has been shown that *patients remember less the one-third of the information given them* by their genetic counselor, with the amount and level of *retained information diminishing even further* if the news presented is either shocking or upsetting.[84]

Genetic screening is currently undertaken in Australia mainly on newborns.[85] In fact, most Australian children are screened not only for phenylketonuria (PKU), but more and more for cystic fibrosis as well.[86] Although of no compulsory nature, these attempts at screening have met with success and with parental approval.[87]

Amniocentesis

Health Commission policy in New South Wales encourages women forty years of age or over and those with a family history of genetic disorders to avail themselves of amniocentesis during their pregnancies. The procedure is only "available" to women between the ages of 35 and 39 and to those who had had a previous child with a disorder which would have been potentially identifiable through amniocentesis.[88] The *practical* application of this policy is to restrict the use of the procedure itself to those women who show through their family history of the previous birth of an abnormal child, or are either 40 years of age or older, that they are "at risk" in their pregnancy.[89]

The availability of amniocentesis varies from state to state, with some authorities even allowing it at will for all women 35 years of age.[90] In Queensland, the procedure's use is unrestricted as to age.[91] The interesting point here is that the costs involved in diagnosing one handicapped fetus (disregarding the reduced parental anxiety) are estimated to be less than one-twelfth the cost of maintaining a resulting abnormal child in a public institution for 10 years.[92] For an average lifetime, it has been estimated by the New South Wales Health Commission that approximately $500,000 will be spent for one institutionalized person born with a genetic abnormality.[93]

THE NEW BIOLOGY AND A PROGRAM FOR POSITIVE EUGENICS

Artificial Insemination

Artificial insemination, referred to as AID or heterologous insemination, is the process of inseminating a woman with the sperm of a donor. Although AID was developed to provide a child to a married couple who could not reproduce due to a physical impediment of the husband, the method today has a new vitality and purpose as a technique for implementing a program of positive eugenics.[94] Sperm banks have been established to maintain semen of "distinguished" men even beyond their lifetime.[95] Positive eugenists advocate superior sperm banks in order to develop the population to a position of genetic strength and to assure the survival of the human race in the event of an insufficient number of acceptable males to allow normal reproduction.[96] The ultimate goal of positive eugenics is to assure *eutelegenesis*, mass insemination with superior sperm.[97]

Interestingly, the very word eutelegenesis was first proposed by Marion Piddington in 1916 "as a means of populating Australia and creating a race combining high moral worth with sound physical development," and was used subsequently by early American eugenists.[98]

The idea or suggestion for use of AID practices to implement a program of positive eugenics should, in theory, encounter little resistance because these practices infringe upon individual rights only minimally, neither restricting nor prohibiting marriage or reproduction.[99] Of course, there are varying ethical and moral issues associated with this practice by unmarried women.[100]

In Vitro Fertilization and Embryo Implants

In 1974, Dr. Douglas Bevis of Leeds University announced that out of approximately 30 attempts to conceive human embryos in vitro, or in test tubes, and then implant them in utero, or into the wombs of women, he had achieved three successful implants that resulted in the births of three babies.[101] The three mothers had been infertile because of diseased, blocked, or missing Fallopian tubes. Dr. Bevis had removed ova from each woman, fertilized them in the test tube with sperm taken from the women's respective husbands, and then implanted the fertilized eggs into the women's wombs.[102] Because he was unwilling to fully document his research, Dr. Bevis's announcement was subject to considerable doubt.[103] It remained for Dr. Patrick Steptoe, a British gynecologist, and Dr. Robert Edwards, a Cambridge University physiologist, to document laboratory conception of a test tube baby and of its birth in 1978.[104]

In Australia, Dr. Carl Wood of Monash University and the Queen Victoria Medical Centre in Melbourne has gained for himself and his country worldwide credit for perfecting and advancing in vitro fertilization techniques, and especially the utilization of frozen embryos as a means of combating infertility.[105] A plethora of vexatious moral, ethical, and religious issues have been raised regarding the status of frozen embryos and are beyond the scope and purpose of the present discussion.[106] What may be acknowledged, however, is the reality of increased use of in vitro fertilization and embryo transplants in humans so long as no other means of conquering infertility are discovered or made available.

If a woman is infertile due to a blocked or missing fallopian tube, an ovum may be taken from one of her ovaries, fertilized in a test tube with her husband's sperm (or a donor's sperm if her husband is himself infertile) and implanted in her uterus. If a woman cannot produce normal egg cells, a donor's egg, already fertilized by the husband's sperm through artificial insemination or fertilized in vitro with the husband's sperm, could be implanted into her uterus.[107] A woman who cannot carry a baby to term because of a physical disability could enter into a contract with a surrogate or host mother to do so,[108] and an egg fertilized either in vitro or in vivo could be implanted into the host mother. A career woman, such as a professional athlete, for example, who has no physical disability, may also seek the services of a surrogate mother if she does not wish to miss valuable time from her professional interests to carry a baby for the full term.[109]

Successful in vitro fertilization also may lead to the development of in vitro gestation, thereby enabling a fetus to develop to term completely outside the womb.[110] Married couples could also rely, additionally, on in vitro fertilization techniques to have a child that was not even genetically their own. And, of course, an unmarried person desiring a child might wish to utilize these methods as well. Since an unmarried individual would need a donor's egg or sperm to effectuate the procedure, such a program could introduce positive eugenic concepts to create children with a stronger genetic heritage.[111] As in the case of AID programs, the incorporation of positive eugenics concepts would infringe individual rights minimally because they neither restrict nor prohibit marriage or reproduction as eugenic programs do generally.

Asexual Reproduction: Cloning and Parthenogenesis

The word *cloning*, which derives from a Greek root meaning *cutting*, is generally defined as asexual propagation and is a common practice to develop new varieties of plants.[112] In 1966, a team of Oxford University biologists, headed by Dr. John Gurdon, announced that they had grown seven frogs from the intestinal cells of tadpoles.[113] What had been routine in the garden now existed for one group of animals: a new organism produced from a single parent.[114] Several steps would be required to clone a human. First, the nucleus of a donor's egg cell would be destroyed. A nucleus from any convenient cell of the person to be cloned would be inserted into the enucleated egg by microsurgical techniques not yet fully developed. The new cell, placed in a nutrient medium, would begin to divide and embryo implantation would follow in approximately 4 to 6 days.[115] The cloned individual would be the identical twin of the person who contributed the body cell.[116] Significantly, the establishment of banks of tissue cultures would permit the production of genetic copies of deceased persons through cloning.

Parthenogenesis, commonly referred to as virgin birth, is another form of asexual reproduction.[117] The

French-American biologist, Jacques Leob, achieved parthenogenesis in sea urchins in 1899.[118] More recently, scientists have reported laboratory parthenogenic experiments for frogs and mice.[119] If this process is perfected for humans, a woman one day may produce the necessary egg cell for conception, jolt the egg by pulling an electric switch or administering a necessary drug, thereby enabling it to split, and then have it implanted in her womb for gestation and ultimate birth—all without physical contact with a man sexually or with his sperm artificially.[120]

Not enough is known, either technically or ethically, about human cloning or parthenogenesis to raise dogmatizing concern whether it should or should not be undertaken.[121] Present standards of medical ethics require that a researcher be reasonably confident about the outcome of his research, that he undertake research for reasonably humanitarian purposes, and that he obtain the informed consent of the research subjects.[122] These factors do not force the conclusion that cloning is or is not admissible. If the rate of pollution of the human gene pool continues to increase through uncontrolled sexual reproduction, however, efforts to produce healthier people may be required to compensate for the spread of various genetic diseases.[123] In that event, one could make a strong ethical argument to justify cloning of healthy individuals on the ground that it could achieve the greatest utilitarian good for the greatest number of people concerned.[124]

Legislation that embodies positive eugenics concepts for permitting only individuals with superior genetic endowments to clone raises a serious constitutional issue. Such a statute would require safeguards against the large-scale cloning of particular types of individuals; to do otherwise would decrease the genetic *variation* that is so vitally necessary to natural selection, and would even threaten man with his own eventual extinction.[125] By discriminating between those with superior genetic traits and all others, however, legislation of this nature would be subject to equal protection challenges. Under standard equal protection analysis, if a court determined that the statutes affected a fundamental right, the state would need to show that the legislation served a compelling state interest by its enactment and enforcement.[126] The right to procreate is, as observed, a fundamental right.[127] But, the denial of cloning methods to individuals who are capable of reproducing in the normal manner may not be a sufficient infringement of this fundamental right to trigger the compelling interest requirement.[128] If it were not such an infringement, the state would be required only to show a rational relation between the legislation and a legitimate state interest.[129] A court might determine that the state's interest in the propagation of superior traits is constitutionally impermissible because it violates the Constitution's nobility clause or the Thirteenth Amendment's prohibition of involuntary servitude.[130] If a court determined that the state has a legitimate interest in the propagation of superior traits, it would probably go on to find that the legislation is rationally related to that purpose.

Persons who carry recessive traits might succeed in claiming that permitting only genetically superior people to clone infringes upon their right to procreate—with that claim thus triggering strict judicial scrutiny of the cloning law and requiring the state to show a compelling interest for its action.[131] Under this type of judicial scrutiny, at least two attacks on a statute itself could be made in addition to challenging the state's purpose for action as constitutionally impermissible. It is doubtful whether scientific evidence can provide a rational basis for classification of individuals have superior genetic traits.[132] Moreover, the state may be able to achieve its objective through a less intrusive program: its interest in the propagation of superior traits through a positive eugenics program is probably less compelling than its interest in the diminution of inferior traits through a negative program.[133]

THE MYSTERY AND THE PROMISE OF SOCIOBIOLOGY

The sociobiology debate has been described as "the continuance of the historic conflict created in the social sciences and humanities by the mechanistic examination of human nature through the instruments of conventional biology."[134] Strictly as a discipline, rather than a theory, sociobiology is defined classically as the "systematic study of the biological basis of all social behavior,"[135] with human sociobiology but one aspect of the whole study of the biological basis of social behavior.[136] Stated otherwise, sociobiology is but the study of the "evolutionary roots of social behavior."[137] Evolutionary sociobiology's goal should be not only to reconstruct the history of primates and identify their course of adaptation over time, but to monitor the genetic basis of current modes of social behavior.[138] As Edward O. Wilson, the modern-day progenitor of Sociobiology, has stated:

> Contemporary general sociobiology might at best explain a tiny fraction of human social behavior in a novel manner. Its full applicability will be settled only by a great deal more imaginative research by both evolutionary biologists and social scientists. In this sense the true creative debate has just begun.[139]

Gene sovereignty or biological determinism[140] remains under constant challenge by environmentalists who assert that, as to sociobiology, there is no genetic variation in the transmission of culture. "Culture," noted Dobzhansky, "is not inherited through genes, it is acquired by learning from other human beings. . . . "[141] Boulding's theory of "Ecodynamics" builds upon a nonbiologic process that posits that each generation of human beings learns more from the last *culturally* rather than by the inheritance of biologically predetermined genes.[142]

The assertions made by sociobiologists that sociobiology allows for an opportunity to explicate heretofore inexplicable behavioral phenomena within a restructured framework of contemporary Darwinian evolutionary theory has rekindled a strong biological interest in the sociobiology discipline.[143] Although there has been substantial criticism about what is perceived as the illegitimate use of biological analogy in analyzing social systems,[144] and the inherent weakness of the nonverifiable assertion of the sociobiologists that the existence of human social structures exists because of a superior adaptive value,[145] the efficacy and relevance of the theory of sociobiology for the study of both human behavior and human nature is of unique significance because of the fact that it "stands as an instance of a rarely observed intellectual phenomenon: the attempt to produce and legitimize a new scientific discipline."[146]

Evolution may be regarded as "a competition for survival among genes,"[147] with the survival depending in large part upon regeneration of the species.[148] This in turn will be tied to a standard of evolutionary behavior which will mandate, all things being equal, a form of altruistic conduct promotive of this regeneration.[149]

> The evolutionary theories of sociobiologists show that
> beings who considered only their own interests would
> leave fewer descendants than beings who also consid-
> ered the interests of their kin. So there is a good reason
> to believe that we do not all act solely in our interests.
> Genes promoting strictly selfish behavior in individual
> animals would be less likely to survive than genes which
> do not.[150]

Relying upon the principle of reciprocity, sociobiologists
suggest that two forms of altruism are at work in the
process of natural selection and propagation of the gene:
kin altruism and reciprocal altruism.[151] Both forms are,
in an ultimate sense, promotive of the "selfish gene's"
best interest of survival and propagation.[152]

The effect of biological evolution upon the development of law has been both studied and evaluated for
quite some time.[153] Indeed, it has been suggested that
the legal "roots" of sociobiology are to be found in the
writings of Maine, Corbin, Wigmore, and Holmes.[154]
The very theory of legal evolution was structured by
Holmes when he observed that

> The life of the law has not been logic it has been expe-
> rience. The felt necessities of the time, the prevalent
> moral and political theories, intuitions of public policy,
> avowed or unconscious, even the prejudices which
> judges share with their fellowmen, have had a good deal
> more to do that the syllogism in determining the rules
> by which men should be governed.[155]

He continued:

> [T]he law is always approaching, and never reaching,
> consistency. It is forever adopting new principles from
> life at one end, and it always retains old ones from his-
> tory at the other. . . . It will become entirely consistent
> only when it ceases to grow.[156]

Efforts in the present day are being undertaken to

postulate a theory of sociobiology for aid-giving actions that have legal consequences[157] and, more especially, intestate wealth transfers,[158] general property rights,[159] privacy,[160] and the doctrine of nuisance.[161] Indeed, although biological theory may offer no unquestioned answers of why certain legal outcomes result from genetic alignments, important partial explanations may be proferred.[162] Nay-sayers do exist, however, and are quick to note that evolution has had little effect on the bulk of law.[163]

CONCLUSIONS

Eugenics clearly enjoys both a positive and a negative relationship with genetics. It has a negative force or potential (as does human life itself), to be sure; but the threatening dimension of its unrestrained application is of minor consequence when the positive sequence of its contributions are charted and realized. The dynamic vectors of force seen in the application of modern eugenics through efforts of genetic advancement and "engineering" must be restrained and placed in equilibrium in order to alleviate fears of unbridled slippery slopes of scientific advancement.[164] Viewed as not only an aid to the tragedy of infertility in family planning, but as a tool for enhancing the health of the nation's future citizens, vital research and experimentation must continue apace in eugenic and genetics. To separate one from the other assures an impotent, as opposed to a virile, response to both the challenge and the mystery of the startling yet controllable development of the new reproductive biology.[165]

Controlled breeding through genetic manipulation is not far behind the legalization of artificial insemination. Once public acceptance of AID is achieved, rapid progress will be made in advancing similar recognition of other new techniques. The law then will be in a better position to begin to chart a course of action and keep pace with science instead of trailing behind to grapple with the scientific, legal, ethical, and social issues of the Brave New World. Although some assert that eugenic control or controlled breeding is dangerous, foolhardy, destructive of the integrity of the family, and violative of the human right to determine the size of the family unit, the unalterable fact is that forecasts indicate that the world soon will be overpopulated if appropriate actions are not taken. Genetic planning and eugenic programming are more rational and human alternatives to population regulation than death by famine and war. Quality of life, in the final analysis, must be recognized as a responsible coordinate to sanctity of creation.[166]

If we approach mastery of the genetic code and seek to acquire a greater understanding of sociobiology as a practical discipline, all with careful resolve to minimize human suffering and maximize the social good (here, the maintenance of health and prevention of disease), we will approach the future with assurance that, as Daedalus, we will in fact arrive safely and meet our goal. If we set out with reckless abandon and are driven only by blind instinct, we will surely be corrupted and, as Icarus, fall.[167]

Chapter 5

Organ Procurement and Transplantation

The Scope of the Problem

INTRODUCTION

The *Times* of London reported on June 11, 1987 that on one of her reelection campaign stops, when approached to volunteer for LIFELINK, a transplant donor register, Mrs. Margaret Thatcher agreed and signed a statement acknowledging that her organs could be removed, upon her death, and that, specifically, her heart, lungs, kidneys, eyes, and skin could be given to patients awaiting transplants.[1] As a registered donor, Mrs. Thatcher became one of 11,000 who have already become participants in a pilot project based in Birmingham that hopes to become a national computer record of all such potential donors. Ideally, all large British hospitals will subscribe to LIFELINK and thus be given direct computer access to the registered donors, so that when a patient dies, hospital authorities will be able to determine immediately if he were registered with LIFELINK.[2]

In practice, the donor card whereby an organ donor records his intent to have his organs transplanted upon death, that in some states in America is made a part of one's driver's license, has not been as effective a mechanism for facilitating organ transactions as was originally hoped. The decedent's family plays a more determinative role in organ donation and procurement that the potential donor himself. Since few people in the United States sign donor cards, the cards are rarely available when the ultimate decision must be made and, furthermore, procurement teams will *not* normally remove an organ without the consent of the family, and this is the case even with a validly executed card. Thus, in practice, the family is the major source of authorization for the removal of organs.[3]

In 1980, there were some 36 heart and liver transplants in the United States; by 1984, 346 heart and 308 liver transplants—and, in 1985, the statistics doubled again. Kidney transplants rose from 4697 in 1980 to 6968 in 1984 and then to 7695 in 1985.[4]

Because of phenomenal strides in transplantation techniques and related technologies—and particularly in immunosuppressive therapies—the problem of organ procurement and transplantation has become acute. Accordingly, as various forms of transplantation become *accepted therapy* for a wide variety of life-threatening conditions, and the capability to perform such surgical interventions becomes increasingly available, the shortage of organs will directly cause a greater number of potentially avertable deaths.[5]

Approximately 20,000 people die in the United States each year under circumstances where their organs could be salvaged to benefit others. The gap between the number of salvageable organs and the number actually salvaged is much greater than the number of kidney transplants in 1985; but 1876 of those kidneys were actually donated by living donors, and most cadavers provide two kidneys. Thus, it is necessary to subtract the number of living donors from the total figure and then divide the number by two. When due allowance is made for the wastage of organs (perhaps 20% in 1985 for an estimate of 1000) and for the 200–250 kidneys shipped abroad, it is reasonable to suppose that there were approximately 3500 donations of cadaver kidneys in 1985. The active waiting list for kidney transplants is estimated to be 9000 to 10,000 and it is likely that many more of the approximately 75,000 patients on kidney dialysis would be good candidates for a kidney transplant, if more kidneys were available.[6]

Signed into law in October 1984, the National Organ Transplant Act addressed several areas of governmental concern, including a prohibition of the sale of organs. The establishment of a Task Force on Organ Transplantation charged with the development of a comprehensive review of medical, legal, ethical, social, and economic issues in transplantation was also created under the legislation; this together with a new Office of Organ Transplantation within the Public Health Service of the Federal Department of Health and Human Services.[7] Under the Omnibus Budget Reconciliation Act of 1986, authorization for financial coverage of immunosuppressive medications for outpatients for 1 year after the transplant was guaranteed.[8] A number of states have undertaken major studies of organ transplantation and some 26 or more have established specific policies for hospital practice within this area.[9]

Within all 50 states and the District of Columbia, the legal framework for organ procurement is found in the Uniform Anatomical Gift Act.[10] Thus, under the provisions of the Act, individuals may determine what will be done with their organs after their deaths, or, in the absence of a valid expression of the decedent's wishes, the family may decide what to do with the organs. In order to eliminate confusion about the disposition of bodily parts, the Act clearly establishes the wishes of the decedent as determinative, while recognizing a role for familial wishes when the decedent's wishes have not been declared or validly expressed.[11] Because of the necessary goodwill that must be maintained by the hospital procurement teams, the reality of the practice—regardless of the Act—as noted, means that a grieving or unsophisticated family member may often veto the wishes of the decedent and the procurement team will respect that action.[12]

A 1985 Gallup Opinion Poll in the United States showed that only about 45% of Americans polled were somewhat likely to donate their organs upon death. There are several reasons for the apparent reluctance to donate one's own organs where such action requires the execution of a donor card. One, people dislike "thinking about dying" and of having their body cut, mutilated, or dismembered upon their death. More important, however, is the fear that, with the donor card, the attending physician might well do "something" before death really occurs or, for that matter, hasten the death itself in order to harvest the donor's organs.[13] It can be seen clearly that the medical profession must go a long way to allay these fears and instill trust among the citizens about the purposes and values of organ transplant programs. Education will obviously take time.

Two Catholic priests have suggested recently that the presumption regarding the disposition of a cadaver should change, with the organs of a dead person being harvestable for reuse *with or without* family consent before the cadaver is ever released by the hospital for burial; the new presumption being that a deceased person would be willing to have his organs assist others if at all possible.[14] The priests structure this position as a consequence of their belief that everyone has an obligation to help the sick who are unable to achieve health by their own efforts and that by allowing a transplant, nothing is really taken away or lost from the decedent but, rather, he and his family gain from rendering assistance so that another may have a better life.[15] This approach will also take a considerable amount of educational packaging before it really "sells."

A SIMPLISTIC PURPOSE

The primary concern of transplantation is either supplementing or replacing mal- or nonfunctioning elements of the body with a naturally or artificially designed one. The effect of such efforts, then, can be seen as either prolonging life or the death process and, furthermore, as augmenting or attenuating the very quality of living or dying.[16] Thus, a substantial connection is seen between "death control" and organ transplantation.[17] With the enhancement of patients' lives as a consequence of transplantation comes an advance in scientific and medical research and its vital clinical application.

At least well into the foreseeable future, even if vast resources were allocated sufficiently to improve transplantation procedures, the major problem would remain that of supply. Of equal concern are problems of logistics, rejection, and storage. Interestingly, if the rejection problems were solved, the supply problem would be exacerbated in one respect by having the effect of increasing the pool of appropriate recipients.[19] And, even with a massive infusion of resources that allowed a resolution of logistical, storage, and rejection problems, the supply problem would continue. It should be realized that this problem of supply is not tied exclusively to economic or technical considerations, but to a very large extent to both moral and legal barriers that limit the number of organs available from live or cadaver sources.[20]

Cadavers

If saving life is to be recognized as the central most ethical principle not only of medicine, but of law and religion as well, then there should be no question that organs from cadavers may be salvaged and used to save the lives of the living.[21] The actual "harvesting" of cadaver organs should be done so as to minimize any traumatic effect of the practice upon grieving relatives and with respect for donors' religious beliefs. Thus, if one were, for example, a Fundamentalist Christian who considered organ removal as inconsistent with the principle of bodily resurrection, he should not (upon his death) be considered as a donor.[22]

Applicable policies should be structured in order to allow the routine removal of useful cadaver organs and allow a prospective donor to object, during his life, to the removal of his organs after death. In those cases where the donor consents to the use of his organs upon death, no family member should be empowered to veto this decision. Where a prospective donor has neither specifically objected nor expressly assented to a removal of his organs at death, the next of kin should be allowed to sustain a controlling objection to the removal any time before the organs are actually harvested. Reccgnizing this principle would serve to obviate any constitutional problems regarding freedom of religion.[23]

As to the issue of who has the burden of action in salvaging the actual cadaver organs—the surgeons seeking consent, the donor before death, or his next of kin objecting—the burden should be shifted from the surgeons to the deceased donor or to his next of kin. He should instead be told: "You may remove cadaver organs to save the life of a living person unless the deceased notified you that he objected or the next of kin now objects."[24] Accordingly, if this were done, the routine removal of cadaver organs would be accomplished unless some specific objection was entered before removal.[25] The burden of action would then be placed upon the person who did *not* wish the organs removed to enter his objection. Adoption of this policy would not only serve to satisfy the medical need for the immediate removal of organs (without undue delay in obtaining consent) but would additionally produce far more kidneys for transplantation, establish a routine procedure at death, and prevent organ removal from bodies of individuals who objected during life or whose next of kin objected after death.[26]

Living Heads

Patrick Kelly (aka Chet Fleming), Esquire, has a unique and unusual patent in number 4,666,325: for it describes a machine to keep alive severed heads, be they animal or human.[27] In 1970 a Rhesus monkey was decapitated and the head was mounted successfully on a mechanical support for some 36 hours.[28] Although the machine under Mr. Kelly's patent has yet to be built, it is regarded as theoretically possible since essentially *all* that is required to make it operable is a heart-lung machine, an artificial kidney, and several other devices to add nutrients and keep the parts under control generally. It has been estimated, however, that it would cost $5 million to build such a machine.[29]

Total head and brain transplantations are, as an idea, as fascinating as they are frightening. Dr. Robert While, a neurosurgeon at the Case Western Reserve University School of Medicine in Cleveland, predicts that as the twenty-first century approaches, more living heads will be preserved either artificially or by transplanting them to another body—and this will be done in humans.[30]

The Neomorts

The neomorts, or "living dead," are brain-dead with, unlike comatose patients, no chance of being revived. Generally, the neomort can exist for upwards to 2 weeks while researchers utilize them to test new medical procedures and to store organs for transplantation.[31] It has been predicted that within a decade, many hospitals will have facilities to house neomorts and begin using them for scientific research.[32] People eventually become neomorts only as a consequence of giving their permission, before death, for the use of their body in this manner or if their families give permission when the patient himself has not otherwise directed a use of his body at death.[33]

The Irreversibly Comatose

This tolerant and practical attitude about neomorts should also be extended to persons suffering irreversible coma who, while they can never regain a state of consciousness, owing to the fact of irreparable damage to the cerebral cortex, still have a certain level of activity in the brain's lower levels. Persons in this state can breathe on their own and live for years, as did Karen Quinlan, with intravenous feeding and nursing care.[34] The costs of maintaining these human vegetables may exceed $25,000 per year, with the use of staff being an additional economic drain. From the standpoint of making an optimal allocation of economic resources, such expenditures are exceedingly hard to justify.[35] Accelerating the death of such human vegetables would automatically be labeled active euthanasia and incur public censure.[36] Yet, in time, the increasing and valid needs of the *living* who, with suitable transplants, are salvageable under the long-standing principle of triage, will simply have to assume a priority of need over the human vegetable.

Xenografts, Anencephalics, and Abortuses

Undoubtedly, however, the two most critical ethical issues to be raised of late involve the use of xenografts from nonhuman sources and the use of anencephalic donors and abortuses as donors.

Given the scarcity of cost of primate sources, is it not medically prudent to begin to develop techniques for transplanting organs from animal sources to human beings? Of course, the attempt to transplant the heart of a baboon into the chest of a human infant at the Loma Linda University Medical Center in California was the event that triggered this high level of concern.[37] Dr. Leonard L. Bailey, who transplanted the baboon heart, stated that the moral validity of his actions was not debatable (as he acted properly) and expressed his intention to ask his hospital to allow him to utilize monkey hearts in dying babies born with defective hearts until they can receive a human heart.[38] The President of the International Society for Heart Transplantation agreed that such transfers should be allowed and treated as experimental in nature.[39]

If human cadaver organs are available for transplantation, it would appear ethically indefensible for animal sources to be used. However, it should be remembered that the shortage of organs and tissues available from human sources for infants is far more severe than the serious shortfall that exists for children and adults. The small size of newborns means that, for the most part, only organs from other infants can be used for transplantation. Thus, the gap between supply and demand is likely to grow greater in the years to come as transplantation techniques for infants are perfected.[40] The only means of treating life-threatening congenital organ failure will be through the use of xenografts— unless, that is, a means can be otherwise developed for increasing the supply of cadaver organs and tissues from newborns and infants.

In the United States each year, approximately one out of every 1,800 births results in a child who is afflicted with a fatal, incurable congenital defect known as anencephaly. These tragic infants are born with all but a small portion of their brain missing and normally survive but from 7 to 14 days.[41] Many, in fact, die in utero.

The laws of a number of state jurisdictions require that death be pronounced using brain-death criteria or the cessation of heart and respiratory functions before the procurement of organs is allowed. Sadly, anencephalic infants meet these requirements. The dilemma here is that if physicians wait until all electrical activity from the small portion of the brain present is extinguished, there is grave concern that its vital organs and tissues will be severely damaged.[42]

The use of fetal tissue from abortuses is a practice shrouded in great secrecy both in the United States and throughout the world. The aborted fetus can serve as the source of various tissues, including nerves, brain tissue, stem cells taken from bone marrow or the developing liver, and cells from the pancreas. Again, in the United States, there are prohibitions on fetal experimentation in utero, but no such regulations apply to the use of tissues and organs derived from cadaver sources.[43]

For the approximately 500 children born each year in the United States who suffer from end stage renal disease, a similar number born each year with congenital defects of the heart and the well over 1000 born each year around the world with congenital liver failure, the anencephalic newborns and abortuses provide a desperate opportunity to live.[44]

SELECTION PROCEDURES AND BASES

Just as there are no uniform sets of criteria for selection of patients for kidney transplantation,[45] so too is this a dilemma for other transplant efforts. The search for rational and principled standards of apportionment of scarce resources is and will remain a vexatious problem for decades to come.[46] Informal "rules of thumb" simply cannot be countenanced.[47] One way to avoid totally the problem of distribution is to avoid using the scarce medical resource altogether. Thus, when one hospital was faced with a curtailment of dialysis services and a corresponding inability to treat but a small portion of its patients, it chose totally to cease offering dialysis treatment.[48] Having exceeded its budget, Guy's Hospital in London simply conserved hospital resources by ceasing operations on heart patients and curtailing kidney dialysis use and intensive care for premature babies.[49] Some might characterize these actions as unreasonable, while others term them but fiscally prudent. Inevitably, the medical screening of patients in order to determine their priority for treatment must be tied to the time-honored principle of triage that states that basically the most salvageable should receive a priority in treatment decisions.[50]

Both in the formation of the transplantation waiting list and the actual distribution of donated organs, there is a consensus that the primary criteria at both stages should be medical: that is, medical need and probability of success. The debate focuses, however, around the issue of whether these medical criteria should be defined broadly or narrowly, about the manner in which they should be specified, about the relevance of such factors as age and life-style (i.e., is the candidate for a transplant an alcoholic who would thus jeopardize the success of the donation) and, finally, about whether need or probability of success should have priority in cases of conflict. If there is no reasonable chance that a donated organ will benefit a particular patient, it is considered unethical stewardship of the actual donated organ to implant it in the at-risk patient.[51]

Debate also focuses on what an unequivocal success will be: length of grant survival, length of patient survival, quality of life subsequent to the transplant and rehabilitation or salvageability, consistent with basic principles of triage.[52] Additional harsh criticism and debate has, of late, turned to policies regarding nonimmigrant aliens. More specifically, should some such aliens be placed on waiting lists and actually have the opportunity to achieve a priority ranking, should aliens be excluded from receiving any cadaver organs obtained in the United States; should numerical quotas be set or should the nonimmigrant alien be accepted for a waiting list but informed that he will not receive an organ until it is first determined that no United States citizen (or alien residing in the United States) could benefit from that organ?[53]

Georgetown University's Hospital in Washington, D.C. is the only one of Washington's transplant hospitals with a stated policy giving its American patients, as well as Americans on waiting lists at other hospitals, preference. Several of the staff surgeons at Georgetown note candidly that this preferential policy is not always followed strictly. Other area transplant surgeons recorded their repugnance at the idea of discriminating among patients on the basis of nationality.[54] The administrator of the University of Minnesota's Hospital, C.E. Schwartz, has observed that some hospitals impose a numerical limit on the number of transplants they undertake within a given time; this being done primarily in order to stave off financial disaster.[55] At his own hospital, Mr. Schwartz related that a deposit requirement on out-of-state residents seeking heart and liver transplants is imposed that means that a deposit of 80% of the actual cost of the operation must be made before admission is granted to the nonresident.[56]

Inequality of Publicity and Allocation

Appeal is often made to the High Court of Public Opinion in order to secure necessary publicity and funding for very tragic transplant applicants. Some—because of journalistic access or interest by the media in "unique" individual cases or because of the social or economic status of the applicant and his family—appear to gain an unfair access to the hearts and pockets of many supporters and thereby step to the front of the line, so to speak, in acquiring the attention of potential donors.[57] Seven-year-old Ronnie DeSillers's plight in needing a fourth liver transplant is a case in point. The national publicity surrounding this tragedy was so intense that President Reagan even made a gift of $1000 in order to help defray the costs of the final transplant; but, sadly, young Ronnie succumbed.[58]

The vast amount of money required for a transplant, hospital care, and related costs can typically be in excess of $200,000.[59] States differ in their willingness to commit funds and general support for transplants. While Maryland has shown a more tolerant and willing attitude toward support here, Virginia has been more conservative and frugal. Recently, the parents of a four-year-old child seeking a liver transplant were forced to make an emergency appeal to the Fourth Circuit United States Court of Appeals to override a state refusal to fund the surgery. Virginia's medical policy allows payment for liver transplants for children under eighteen years of age who are diagnosed as having extrahepatic biliary atresia—a somewhat common liver disease. The 4-year-old's rare disease, secondary biliry cirrhosis of the liver, was not covered under state policy.[60] When surgery was completed, the transplant team of four physicians had used 50 pints of blood in an operation that lasted 25 hours.[61]

Gifts

Ideally, society should be able to rely upon altruism to provide an adequate supply of organs for purposes of transplantation and thus eschew all forms of commerce in human organs. Sadly, such an ideal society does not exist and procedures must be developed and implemented in order to prevent undue suffering or premature death.[62]

Fostering the gift relationship and allowing sales of organs might initially be regarded as a solution to problems of allocation and distribution, since organs would be drawn to a pool of available goods by gifts and sales and, at the same time, be distributed to respective donees and buyers. Chance is too dominant an element in random gifts and sales are, for some individuals, a highly questionable way by which to distribute such vital resources, simply because many persons are unable to afford them. The poor would be under "economic coercion" to sell their organs and fear the growth of a "cannibalization" ethic. Perhaps it might be more feasible for gifts to be funneled into a type of general pool and thereupon distributed on some basis other than a donor's specific choice of a recipient. Similarly, private or governmental groups could well purchase parts, organs, and tissues for later distribution by sale *or* gift.[63]

A DISTRIBUTIONAL STANDARD

Distributing scarce medical resources thus involves obvious problems of distributive justice. Although acknowledged as existing, they are quite difficult to resolve in a pragmatic manner. Consequently, owing to this often insurmountable difficulty, the question of how the distribution will be made is reduced to the issue of who will make the first order decision. Yet, unless triage decisions are to be recognized as arbitrary and capricious, some criteria must be in place for scrutiny and examination. The Hemodialysis Program of Seattle

Washington's Artificial Kidney Center studied 87 such centers around the country in order to develop a set of criteria for allowing patients to be admitted to their programs. The dialysis candidate profile that emerged found the following criteria to be used always in the selection and admissions process: medical suitability (good prognosis with dialysis); absence of other disabling disease; intelligence (as related to understanding treatment); likelihood of vocational rehabilitation; age; primacy of application for available vacancy in the hemodialysis program; and a positive psychiatric evaluation (re acceptance of disease and goals of the actual treatment). The following conditions were judged as excluding selection of a patient for participation in the program: mental deficiency; poor family environment; criminal record; indigency; poor employment record; lack of transportation; and lack of state residency.[64]

Fault may be found with one or more of these factors used in selection. But, absent a unifying philosophy of medicine that defines with precision its goals for achievement, acknowledges whether such achievement is possible, and determines whether it reflects a desirable goal of contemporary human culture and develops rational guidelines for making necessary critical choices, medicine will not be successful. That medicine has existed in the twentieth century without a vital philosophy is due to the simple fact that its successes "in curing" have been so enormous and overpowering. Indeed, "this success is due to knowledge and technique based on experience, not theories and philosophical speculation."[65]

Today, there is a recognition that an admirable goal of a national health policy is quality health care at an affordable cost. Cost containment thus has become a major force of wide significance and application in all levels of health care decision making. There is little disputation of the fact that resources are scarce relative to wants and that they have alternative uses; and furthermore that differences in individual wants mean an assignment of different values to these wants. The basic dilemma, then, is where to determine a line of compromise between competing interest groups.[66]

Principles of Allocation: Utilitarian versus Egalitarian

Since the law provides at present no uniformly agreed-upon principles that may be applied in order to regulate the allocation of scarce medical resources, current medical practice draws upon a structure for decision making evolved as such from a number of philosophical and ethical constructs. There are five utilitarian principles of application that are operative in the hierarchy of triage: the principles of medical success, immediate usefulness, conservation, parental role, and general social value. Translated as such into decisional operatives, there emerges a recognition that priority of selection for use of a scarce medical resource should be accorded those for whom treatment has the highest probability of medical success, would be most useful under the immediate circumstances; to those candidates for use who require proportionally smaller amounts of the particular resource, those having the largest responsibilities to dependents or those believed to have the greatest actual or potential general social worth. The utilitarian goal is, simply stated, to achieve the highest possible amount of some good or resource. Thus, utilitarian principles are also commonly referred to as "good maximizing straegies."[67]

Egalitarian alternatives, on the other hand, seek either a basic maintenance or a restoration of equality for persons in need of a particular scarce resource. There are five basic principles utilized here: (1) the principle of saving no one; thus priority is given no one because, simply, none should be saved if not all can be saved; (2) the principle of medical neediness under which priority is accorded those determined to be the medically neediest; (3) the principle of general neediness which allows priority to be given to the most helpless or generally neediest; (4) the principle of queuing, where priority is given to those individuals who arrive first; and, lastly, (5) the principles of random selection, where priority of selection is given to those selected by pure chance.[68]

To the utilitarian, maximizing utility, and hence what is diffusely referred to as the "general welfare," are both the primary ground and subject of all judgments. That which is required in order to maximize utility overall may thus infringe upon an individual's own entitlement of rights to particular goods. Accordingly, moral rights are either rejected generally, or recognized as certainly not absolute.[69]

Philosophy and religion may well provide us all with the necessary balance and direction for life and allow us to develop an ethic for daily living and a faith as to the future; but in cases of neonatology, where law, science, medicine, and religion interact, great care must be exercised in order to prevent inexplicable fears and emotions—oftentimes fanned by journalistic prophets of the "what if" shock culture—taking hold of and thereby blocking powers of rationality and humanness. The basic challenge of modern medicine should be, simply, to seek, promote, and maintain a level of real— and, when the case may indicate, potential—achievement for its user-patients which allows for full and purposeful living. Indeed, man himself should seek to pursue decision-making responsibilities and exercise autonomy in a rational manner and guided by a spirit of humanism. He should seek, further, to minimize human suffering and maximize the social good. Defining the extent and application of the social good will vary with the situation of each case, obviously.[70]

Rules of Exclusion and Final Selection

Perhaps utilization of a Rule of Exclusion might go far to eliminate what may be viewed as the harshness of triage. Under such a rule, some individuals would be simply eliminated from "competition" for the particular scarce modality of treatment or care facilities even if the resource(s) were in unlimited supply. Thus, applying this rule, the scarcity of the resource(s) in question would never even be considered.[71]

Rules of Exclusion are preferable, in certain definite ways, to Rules of Final Selection when implementing the principles of triage. With Rules of Exclusion it is generally unnecessary to make comparisons between specific individuals, for either the patient meets the minimum medical criteria or he does not. When operable, these rules have the appearance of greater objectivity and less arbitrariness than a Final Selection Rule that states simply: "First come, first served." If the standard of exclusion is structured in such a manner and at a level high enough to achieve the purpose of initialy reducing the applicant group to that specific treatment number, the very selection process will turn on the decision of exclusion and obviate the need to even be forced to apply additional rules of ultimate or final selection.[72]

There are, essentially, two approaches to structuring and applying Rules of Final Selection: utilize a comparative analysis of the social utility of curing various patients in a selection pool, or undertake no comparison but rather apply an arbitrary yet egalitarian formula, normally, first come, first served (regardless of whether the first served might be a socially irresponsible derelict).[73]

As observed, medical providers themselves failed in the past to articulate precise rules to guide them in determining patient social utility vis-à-vis use of a scarce resource or, for that matter, to structure a list of exceptions to the first-come rule of final selection. These rules of final selection based, it is seen, on value judgments and value judgments alone, are not arguably within a special area of competence for a physician to make. Contrariwise, rules of exclusion are based on and, indeed, formulated from professional evaluations and considerations and are regarded as less subjective and arbitrary and more acceptable to both patients and doctors alike than the rules of final selection.[74]

CONCLUSIONS

No principle of preference is clearly correct, humane, or totally just. Other suggestions include selection of a patient-user by chance or randomization and queuing, the establishment of separate waiting lists for patients in different age groups and for those with or without families, and, perhaps most ideally, widespread support and development of a program calling for the total utilization of artificial organs which would alleviate the scarcity of natural organs. To one degree or other, all of these suggestions are attractive. Obviously no definite solutions can be submitted here. If, however, health care providers seek to pursue their decision-making responsibilities in a rational manner and guided by a spirit of humanism which minimizes human suffering and maximizes the social good of each situation, a humane standard of justice will be achieved and triage will operate as a complement to its attainment.[75]

Chapter 6

El Dorado and the Promise of Cryonic Suspension

In the autumn of 1983, at Johns Hopkins University Medical School in Baltimore, Maryland, a group of doctors, by lowering the body temperature of a cancer patient 32 degrees from the usual 98.6 degrees for 40 minutes, stopping his heartbeat, and inducing a state of hypothermia approximating suspended animation, while performing surgery to remove a kidney growth which had spread through the vena cava into his heart, unwittingly advanced the possibility for medical science, at some time in the future, to achieve a total body suspension in order to combat physical degeneration caused by such occurrences as cancer, heart disease, and a plethora of other debilitating or fatal diseases.[1] The implications of this process have not only intrigued the medical-scientific community, but have also touched the popular imagination with the distinct possibility of making the dream of immortality by holding illness at bay more tangible, if perhaps not a reality.

Popular interest in cryonic suspension, or "deep-freeze" burial, was highlighted recently in a news story that reported a jury award of $928,594 in damages for breach of contract and fraud against a cryotorium, or place where the suspension of the cryon is conducted, for its failure to provide continuous suspension of two "dead" individuals. The cryons were thawed inadvertently and their family pursued this legal action.[2] The NBC television network, in a segment of its program *Prime Time Saturday*, broadcast on March 15, 1980, reported on the state of the art of cryonic suspension and found that approximately 100 persons had contracted to be frozen upon death, for an initial cost of $12,000 and an annual charge of $2,000 a year for maintenance thereafter. Another figure sets the cost of suspension at $60,000.[3] In 1976, some 24 bodies, or cryons, were then in suspension.[4] The interesting point here is that while people are intrigued by the possibilities and hopeful that cryonic suspension may well yield a solution to the curse of mortality, none are willing to allow the process to be initiated until *after* normal death has occurred.

For the lawyer, the challenge of shaping and developing new legal mechanisms in order to reflect the consequences of cryonic suspension is formidable. Owing to the embryonic scientific stage of development of cryobiology and cryonics, it is difficult for both law and society to interact with uniform purpose and resolve. If the present social temperament or level of awareness is itself recognized as being in flux, the law, as but a reflection of that temperament or attitude, cannot in reality be expected to be bold and decisive. Society's goal of immortality is thus fraught with a number of vexatious conundrums which must be addressed.

THE HISTORICAL PERSPECTIVE

The writings of Hippocrates discuss the control of hemorrhage by use of local cold and, during the Napoleonic Wars, medical literature records successful instances of local hypothermia designed to ease and deaden pain when amputations were performed.[5] Hypothermia procedures were refined after World War II by a French scientist, Dr. Henri Laborit, and used extensively for the treatment of shock.[6] In 1946 the French biologist, Dr. Jean Rostand, successfully preserved frog sperm in a partly frozen solution, followed in 1948 by the work of an English scientist, Dr. Audrey U. Smith, who preserved fowl sperm at low temperatures and used it subsequently for fertilization.[7] Yet, modern scientists concluded that the suspended animation of individual cells was not possible.[8] Whether an entire state of human suspended animation will be achieved is viewed as speculative by many.[9]

Working with low temperature experiments in the 1950s, biologists designed the term "cryobiology" to describe those investigations that were conducted well below normal body temperatures.[10] Cryogenics, then, refers broadly to the technology of low temperature experiments, while cryonics pertains to all disciplines and programs centered on human cold storage.[11]

Well before the 1950s—in fact in 1663—an English scientist, Henry Power, composed a mixture of ice and salt and immersed a jar of eels in it, thereby freezing them. After one night they were revived and the phenomenon known as "suspended animation" was originated.[12] The late 1950s witnessed the most significant breakthrough in hypothermia with the development of the heart-lung machine that allowed chest cavity surgery to be performed by permitting blood to be removed from the body to a heart-lung machine and then pumped through a hypothermia unit in order to cool it before returning it to the body.[13]

Today, a survey of the literature of cryobiology is replete with notable successes in the freeze-preservation of viable cell suspensions, blood serum, and microorganisms, semen and nonviable tissues used for transplantation, cryosurgery, and the preservation of large mammalian organs.[14] Although the experimentation and successes in transplanting human organs proceeds with definite success,[15] a total cryonic suspension of an entire human body *and its revival* has yet to be achieved.[16]

The first freezing or cryonic suspension of a human took place for a Dr. Harold Greene after his death on January 12, 1967, with the whole process of perfusing him taking some 4 hours.[17] The greatest danger for Dr. Greene, as for any person undergoing cryonic suspension, was the need to provide as much expeditious care as possible in order to protect the brain and the cells. The brain remains intact from 3 to 5 minutes, at normal body temperatures, after death. With decreases in the body temperature, the brain can remain without oxygen for an even longer period of time down to $-196°$ centigrade. It is at this degree of temperature that all change virtually stabilizes, and the body may, for an indefinite time, remain in a near-perfect state of preservation.[18]

Because the human body is composed of 75% of its weight in water, and water expands when frozen, the body cells would, if left unprotected, burst upon freezing. Therefore, perfusion is the preferred method of choice for internment rather than embalming for a successful cryonic suspension. Prevention of ice crystals inside the body cells is the basic purpose of perfusion. A protective chemical, glycerol, is combined with dimethyl sulfoxide (DMSO), which serves as a penetrant in carrying the glycerol to the cells through the bloodstream. Consequently, an absorption rate of 90% of the cells' water is achieved. This combination assures that the formation of ice crystals will occur not inside the cell, but outside.[19] Since perfusates with a high percentage of glycerol or DMSO are acknowledged to be toxic to the cells, other chemicals must be used in the perfusion process.[20]

THE CRYONICS MOVEMENT

If the message of life is to live truthfully and thereby accept the fact that life ends at some point for everyone,[21] the cryonics movement smacks of absurdity and whimsy. If there is the remotest chance of succeeding, should not the effort be undertaken, then, in the name of scientific inquiry alone, to explore those chances? Unlike DNA experimentation, no real or substantial harm inheres to society as a consequence of research and experimentation in cryonic suspension. Shattered individual *hopes* are the only fatality in cases of this nature. The power and magnitude of scientific thought and discovery can never be underestimated.

It has been speculated that the desire for future life is owing to a perception that most lives are, for one reason or other, incomplete and from a desire to renew friendships which have ended prematurely.[22] While the motivation of the cryonics movement may well be acknowledged to be a conscious or an unconscious desire for immortality, the movement cannot be separated totally from a sustained effort by society as a whole over recent years to prolong life.

For the modern immortalist, the pathway to his goal begins in a freezer. After death, cryonic suspension is administered, and the body is frozen and stored at either the temperature of liquid nitrogen or liquid helium until medical scientific advances are such that the incurable illness that brought about death has been conquered and new life may be assumed. Thereupon, the cryonically suspended individual is taken from his container/coffin, thawed, revived, repaired, and given new life.[23]

It has been noted that "doubt concerning one's future after death represents a state of mind which is practically unbearable for anybody."[24] The Permanent Revolution Against Death—which actively began with Socrates—will have succeeded, once a cryonically suspended individual is revived and rehabilitated.[26] The work of the "Revolution" will be complete when unavoidable damage done to cells and tissues while the body is in suspension is repaired by methods now unknown, successful techniques for thawing have been established, remedies for the terminal illness or deteriorative effects that ended a cryonic patient's life have been discovered, and efforts to arrest, stabilize, or reverse the aging process have met with success.[27]

In 1965, the first cryonic society was established in New York, and in 1966, a Life Extension Society Conference was held in Washington, D.C. The impetus for this activity was the publication of Robert Ettinger's book, *The Prospect of Immortality*, in 1964.[28] Cryonics appeals to humanists, Christians, and Jews alike. Indeed, since both Christianity and Judaism are "life-affirming" religions, the initial clerical reaction to the movement as a whole has been of a friendly nature,[29] with encouragement from some clergy for more serious work to be undertaken in the field.[30] The fundamental teaching and acknowledgment of the Christian faith that the "resurrection of the body" and eternal life are the ultimate hope and salvation is not thought to be in conflict with the cryonics movement. Freezing or suspension is not considered to be a scientific resurrection without an ultimate day of judgment. It is viewed as an extension of life. And extensions of life would enable individuals to continue to follow the ethical and moral codes of their faiths.[31]

The cryonics movement does not appear to be growing. The movement's failure may be due to lack of a charismatic leader. To be classified as an active movement or force, there must be a leader possessed of considerable magnetism, a tangible track record of some success, and a shared philosophy. Robert Ettinger, the progenitor of the movement, is regarded as an "unassuming, middle-aged physician professor, an intellectual and idealist who is inspired rather than inspiring.[32] There is no recorded success of a cryonic suspension and revivification,[33] as observed and, for some, the very promise of immortality inextricably tied to the movement is "in actuality, a threat to one's peace of mind."[34] The escalating costs of preparing and sustaining the suspension process itself precludes a strong enrollment in the ranks of the movement.[35] The growing absence of a skilled and professionally competent organization to maintain the suspension process is also of negative import to would-be cryonicists.[36]

Another obstacle to the movement's success is current public opinion against those who espouse a radical philosophy of self-interest.[37] Within the membership ranks of the cryonic societies can be found a pervasive lethargy. Membership in a group which offers little social activity or neglects to structure a rewards system for its members seeking to postpone all forms of individual and group gratification until death, obviously has a tremendous "current interest" obstacle to overcome in order to maintain contemporary vitality.[38] Until the time when the first cryonaut returns from his frozen habitat, no validity or efficacy will inure to what may be loosely termed the cryonics "movement." Even beyond a successful revival, the cryonaut or cryon will face serious problems concerning social and economic adaption in a society where family and friends are dead, and the indicators of economic wealth have changed dramatically. Thus, the personal problems that he faces become the problems of the whole movement itself.

The message of the cryonics movement should not be seen as a shallow, unsophisticated philosophy of hope bereft of an organized and rational form of operation, but rather the message should be viewed as a call to expand our sights and vision regarding gerontology and the aging process. Viewed within a context of this nature, the movement becomes less a group of frenzied immortalists and more a group of concerned and devoted individuals seeking to learn the message of death through active, healthful living.

CHALLENGES TO LAW AND MEDICINE

The major concern of both law and medicine in meeting the challenges presented by the developing use and eventual perfection of cryonic suspension is to organize itself in such a manner as to perform a full and active partnership in this area where decision making is demanded by society for its own long-range interests. Law must be not merely anticipatory to the legal challenges of the New Biology, but must develop its basic postulates for action from, by, through, and with medicine.[39] Succinctly, the New Biology is considered to include those biological technologies that allow a rather startling degree of influence over, as well as knowledge about, not only human characteristics (e.g., eugenic predetermination and genetic selectivity) but life processes as well, by so doing, generate major value conflicts and give rise to complex, social, legal, ethical, and religious problems.[40]

A pivotal issue or question concerning the use and administration of a cryonic suspension process is the extent to which a physician may be guilty of malpractice. Moe particularly, the immediate challenge here is the need to clarify the legal-medical definition of death and, where necessary, validate a new definition of cryonic suspension thus avoiding criminal liability for murder. Such definitions would also modify the laws of inheritance. Obviously of concern is the plight of the individual who, trusting that modern medical science will develop a solution to his particular malady, goes to his grave, commits a significant sum of money from his estate for his cryonic suspension, and then expects to be revived in due course and cured of his deadly illness.

Although no reported cases exist where individuals made contracts for cryonic suspension prior to their death only to have those arrangements ignored upon actual death, if situations of this nature were to evolve, several difficulties would be presented. First, in determining the validity of the contract, a court could well hold that the very essence of the agreement was offensive to public morals (e.g., a negation or refutation of death) and as such void, or at least unenforceable. In the alternative, a court could find a contract for cryonic suspension to be valid and one of personal service to the deceased and thus unenforceable once the primary contracting party is deceased. If not regarded as a personal service contract, a court could find the contract valid, and assess damages (difficult though they would be to calculate) in favor of the decedent's estate. Of course, neither of these judicial attitudes would have any practical effect on the decedent, for if the process of suspension is not undertaken immediately upon death, it is of little effect. Any damages assessed for a breach would enhance a decedent's estate, but be of no avail to his efforts to gain immortality through cryonic suspension.

The Phenomenon of Death

Although attempts to draw sharp distinctions between the legal and medical definitions of death have been attempted by serious scholars,[41] the law generally treats the determination as one of fact, made accordingly by the "ordinary standards of medical practice" in each community and guided by the customs and laws of each state.[42]

While not regarded as infallible, the standardized methods for determining death are: irreversible cessation of spontaneous circulation and/or respiration; absence of reflex in the eyes' pupils; absence of brain activity; and, absence of response to nerve stimulations.[43] As scientific advances continue, it may be expected that new criteria will be developed or a finer level of sophisticated application will be achieved in charting the occurrence of death. Owing to the rapid expansion of the technology of biomedicine, it would be unwise for a statutory definition of death to be recognized which would structure criteria for diagnosing time of death for, all too often, the motivating forces behind the drive to evolve a uniform or statutory definition of death have been made by those wishing to ensure a ready source for human transplantation.[44] Trafficking in human body parts, in turn, presents another area of ethical and social problems that has yet to be dealt with by definitive, controlling legislation at the state and federal levels of government.

Meeting in Australia in 1968, the World Medical Association put forth the argument against the use of a precise statutory definition of death by noting: "This definition (of the time of death) will be based on a clinical judgment supplemented if necessary by a number of diagnostic aids (of which the electroencephalograph is currently the most helpful). However, no single technical criterion is entirely satisfactory in the present state of medicine, nor can any one technological procedure be substituted for the overall judgment of the physician."[45]

In 1981, The President's Commission for the Study of Ethical Problems in Medicine and Biomedical and Behavioral Research gave its unanimous approval, in drafting a Uniform Determination of Death Act, that death be redefined as an occurrence where: "(1) irreversible cessation of circulatory and respiratory functions, or (2) irreversible cessation of all functions of the entire brain, including the brain stem. . . . A determination of death must be made in accordance with accepted medical standards."[46]

A New Medico-Legal Definition

None of the current movement in clarifying the legal and medical concepts of death is particularly heartening to either individuals presently in cryonic suspension or those anticipating its use. If one were "suspended" *before* death, the real issue becomes how should the law deal with this occurrence, especially from the standpoint of the disposition of a "decedent's" estate. A working definition of cryonic suspension would thus go far toward easing potential difficulties in this field. Accordingly, cryonic suspension should be recognized and defined in law and in medicine as that state where, under medical supervision, body temperature is lowered to such a degree that a condition of temporary cessation of vital processes is achieved.[47] Given this definition, the vexatious Rule Against Perpetuities might not be a total bar to disposition of an estate.

Legal Complexities in Estate Planning

The Rule Against Perpetuities states that "no interest is good unless it must vest, if at all, no later than twenty-one years after some life in being at the creation of interest."[48] Its object, as first formulated and applied, is the same today: namely, to confine the vesting of contingent estate to a relatively short period after their creation.[49]

Since the cryonic suspension and revival process will probably extend over a number of generations, it would seem obvious that the Rule would be violated. Yet an argument could be made that a cryon could remain in a state of cryonic suspension 21 years without being pronounced dead. At the conclusion of this period, a judicial determination of whether or not a scientific breakthrough existed for a cure of the disease that befell the cryon had been made or was imminent would issue. If one did in fact exist or was predictably in the process of being perfected, an additional period of time (e.g., a 5- to 10-year period) could be arguably allowed for the suspension to be continued and revivification completed. If, contrariwise, no such medical or scientific breakthrough had been achieved or was ascertainable in the immediate future, a final legal determination of the cryon's "death" would be made and the estate settled.[50]

Preventing Murder

In order to allow or even encourage physician-scientists or lay persons to participate in the preparation of an individual for cryonic suspension *before* death, an exculpatory clause in the contract for suspension would have to be inserted which would have the effect of conferring an immunity from civil and criminal liability on the doctors, scientists, and others for either failure to find a cure for the illness of those suspended during the period of suspension or for participating in or supervising a surgical intervention (that is, the initial suspension itself) determined subsequently by a court to be life-ending. It would be wise, furthermore, to have either a judicial recognition of the immunity from suit from a criminal prosecution for murder in connection with the acts of cryonic suspension undertaken by a physician on a living individual or, a state statute, for that matter, which would admit as an absolute bar or total defense the acts undertaken to initiate the suspension.

Presently, to undergo a cryonic suspension one must first be pronounced dead; and once such a pronouncement is made, in order to pay off a life insurance policy (since the policy is actually a death benefit), the insurance company needs a death certificate. If a legal and medical state of cryonic suspension were recognized, a suspension certificate might be issued and the problems here of life insurance coverage would be resolved. It is obvious that the proceeds from the policy would be used to meet the initial expenses associated with the suspension process and its maintenance over the years until revival.

In those cases where, *after* a determination of death is made, one seeks to have his remains cryonically preserved, the law should be more flexible than in the cases where the suspension was undertaken *before* death. Indeed, to fail to recognize death as death would play havoc not only with the law of property and succession, but act to destabilize the very social and religious fabric of society.

CONCLUSIONS: A NEEDED PARTNERSHIP

Rather than wait until the reality of human cryonic suspension occurs in order to map a response strategy or actual mechanism, law and medicine should begin to anticipate and to plan now for this and the other rapid developments of the New Biology and of the brave, yet necessarily somewhat frightened, new world which will come in its aftermath.[51] Only with a full and committed partnership on these issues between law and medicine can enduring progress—as opposed to unchartered chaos—be recorded as the benchmark of the twenty-first century.

Law and medicine can be, however, only as strong and directive as the prevailing standards of morality and social recognition allow. Jaded skepticism is undoubtedly tied to the present state of the art of cryonics. Thus, bold and decisive posturing by law, science, and medicine can, in reality, but find itself *reacting* to scientific advances in cryonics instead of either influencing or directing them. Because of the social realities of the day, then, the prized goal of immortality remains fascinating and intriguing—but, for the foreseeable future, unattainable.

Chapter 7

AIDS

The Private and the Public Dilemmas

The blight of infectious disease has been an all-too-frequent and tragic element in the record of civilization, with plagues recorded in the Old Testament's Book of Exodus 9:14 and as early as 500 B.C.[1] When, in Europe, from 1348–1350 the Black Death or bubonic plague occurred, it was regarded universally as the most devastating and lethal disaster in the annals of recorded history.[2] In 1918, some 500,000 Americans died as the result of a swine flu epidemic[3]; and in 1976, anticipating yet another such epidemic, the federal government initiated an intensive campaign to protect, through free vaccination, those deemed to be within the high-risk group.[4]

The parameters of the AIDS catastrophe today are widening each month, as new developments and statistical records change. A pervasive fear sometimes bordering on hysteria is perhaps the only constant in this equation. A statistical profile is helpful to show the scope and depth of this epidemic as it proceeds to invade every element of the social fabric of life in America and, indeed, the world.

1. The Centers for Disease Control reported in 1987 that 33,997 people had contracted AIDS in the United States since 1981 and 19,658 of them had died. Of all cases, 31,223 were men, 2295 were women and 479 children.[5] Two people die each day in California of AIDS.[6]

2. In 1988, it was reported that there were 35,980 AIDS cases in the United States.[7]

3. The number of infected people is doubling every 14 or 15 months—which would mean at least 2000 people are infected every day.[8]

4. Public Health Service statistics reveal that by 1991 there will be 179,000 deaths from AIDS and 270,000 active cases.[9]

5. Approximately 60% of male homosexual volunteers are seropositive, as are 80% of hemophiliacs.[10]

6. The 10,000 cases diagnosed as AIDS in May 1985 totaled 1,677,900 days in the hospital and expenditures of $1.4 billion before death came.[11] By 1991, the projected outlay in costs for AIDS patients will be $2 billion.[12]

7. The average age at death from AIDS is 35 years.[13]

8. Because AIDS has infiltrated the heterosexual population, a meteoric rise in reported cases of HTLV-III infection is expected because of false assumptions that AIDS is exclusively a homosexual disease.[14]

9. Presently, control of the AIDS epidemic relies upon voluntary measures encouraged by vigorous and widespread counseling and education.[15]

Worldwide, it has been estimated that between 5 and 10 million people have been infected with the AIDS virus.[16] Since it is difficult to obtain accurate information regarding the sexual habits of the population at large, it is equally difficult to map a policy of geographic containment or to estimate accurately the course of the epidemic.[17] One French researcher has found a newly discovered type of AIDs virus spreading across Europe and in parts of South America and has cautioned that it may occur in the rest of the Americas.[18] Indeed, the United States Surgeon General has expressed his concern that there could soon be an "explosion" of the AIDS virus within the heterosexual comminity.[19] And, Professor Stephen J. Gould, a Harvard University biologist, has warned that the AIDS epidemic could well carry off a quarter of the United States population.[20]

DISCRIMINATION IN THE WORKPLACE

Labeled human T-lymphotropic virus, type III (HTLV-III), the AIDS virus attacks the immune system within one's body and thus renders it incapable of dealing with "opportunistic" diseases such as pneumonia and skin cancer. Transmitted as a consequence of sexual contact, blood transfusions, and the shared needles of drug users, the virus may be passed, tragically, to a child from his infected mother either immediately before, during, or after birth.[21]

Since the AIDS carriers develop certain identifiable antibodies to fight the virus, two tests—the ELISA (enzyme-linked immunosorbent assay) and the Western Blot—can detect these antibodies. It should be stressed, however, that even though the presence of the antibodies may be detected, this does not in turn guarantee that the carrier will indeed come down with AIDS—or, for that matter, that the carrier's body continues to retain the virus.[22] The incubatory period, or time between the point of infection with the virus and the onset of the symptoms, may vary from 6 months to 5 years or more. The symptoms associated with the AIDS disease include fever, fatigue, diarrhea, and weight loss together with swollen lymph nodes.[23] The AIDS victims's immune system, in turn, becomes so weakened that opportunistic diseases invade the body and cause death. The death itself is not from the AIDS virus, but rather from the other diseases it spawns.[24]

Since mainstream literature acknowledges that AIDS is spread sexually and by exchange of blood and blood products, the principal avenues for preventing transmission of the virus would be the screening of all donated blood and the education of the populace to modify high-risk sexual behavior (follow "safe sex" practices by using condoms and avoiding anal intercourse) and curtail or limit intravenous drug abuse by use of clean syringes and needles.[25] Thus, casual contact with an infected AIDS coworker brings virtually no risk of contacting the virus to the actual workplace environment.[26] Yet, despite public assurances of this, a near-hysteria may be found in some quarters over the nature and transmission of the AIDS virus.[27] Los Angeles, California, was the first major American community to enact an ordinance to prohibit descrimination in the working environment of those "suffering from the medical condition AIDS whether real or imaginary."[28]

AIDS as a Handicap

The scope of an employer's obligations to an AIDS worker-victim has, until the United States Supreme

Court's decision in School Board of Nassau County, Florida v. Arline,[29] been guided federally by a memorandum-opinion issued by the Assistant Attorney General of the United States in the Department of Justice in June 1986, on the application of Section 504 of the Rehabilitation Act to persons having the AIDS virus.[30] Responding to an inquiry from the Civil Rights Office of the Department of Health and Human Services regarding the scope of the federal provisions against discrimination against handicapped individuals in those programs either conducted or funded by federal agencies, the Assistant Attorney General found that while the *disabling effects* of AIDS or related AIDS complexes may present a "substantial limitation on major life activities," and thus be a handicap, *the ability to transmit the disease to others* is not to be regarded as a protected characteristic under Section 504 of the Rehabilitation Act.[31] Accordingly, an employer could discriminate lawfully against individuals with AIDS if he did so as a consequence of fear of contagion—even though the fear might not be rational—instead of the disability aspects of the disease. No protection would be extended to individuals who test positive for antibodies to the AIDS virus, but are not afflicted presently by the disabling effects of AIDS or related complexes.[32]

The issue of whether contagious diseases are covered by Section 504 of the Rehabilitation Act had, up to this time, never been addressed directly: only inferentially in New York Assn. for Retarded Children v. Carey where the United States Court of Appeals for the Second Circuit concluded that retarded children who were infected with the hepatitis-B virus could not be segregated unless it was shown that a significant risk of infection existed for others.[33]

In *Arline*,[34] the United States Supreme Court raised,

peripherally, some questions regarding the reasoning employed by the Assistant Attorney General in his AIDS-Discrimination Memorandum.[35] Dubbed by some as more a civil rights case than one involving public health,[36] the court held that a black elementary school teacher (here, Ms. Arline) with a history of tuberculosis, could be terminated from her teaching position only if, after a full review of all the medical evidence, it was established that her disease was contagious to others in the immediate working environment.[37] Ms. Arline had contracted tuberculosis at the age of 14 and had been in remission for 20 years until, in 1977, and two subsequent occasions, when she tested positive for the disease. At the conclusion of the 1978–1979 school year, the School Board of Nassau County, Florida, dismissed her because of the tuberculosis. The trial court, while acknowledging Ms. Arline suffered a handicap, found her handicap was not protected by the Rehabilitation Act and the district court concluded Congress never intended for contagious diseases to be included within the definition of a handicap. The Eleventh Circuit Court of Appeal proceeded to reverse, holding that diseases are not excluded from coverage under Section 504 simply because they are contagious.[38]

Chief Justice Rehnquist and Justice Scalia dissented, with the Chief finding that obligations imposed on federal fund recipients must be expressed by Congress in an unambiguous form and that the work of Congress in the passage of the Act was silent on the major question of the case: namely, whether discrimination on the basis of contagiousness constitutes handicap discrimination and, thus, the protections of Section 504 of the Federal Rehabilitation Act did not extend to individuals such as Ms. Arline.[39]

In Footnote 7 of the majority opinion, Mr. Justice

Brennan rejected the arguments made by the Solicitor General of the United States that discrimination based upon legitimate concerns about contagiousness—and particularly for AIDS patients—cannot be viewed as discrimination based upon a handicap, because the disease not only creates a physical impairment but contagiousness as well. Justice Brennan expressly reserved the question "whether a carrier of a contagious disease such as AIDS could be considered to have a physical impairment, or whether such a person could be considered, solely on the basis of contagiousness, a handicapped person as defined by the Act."[40]

Ms. Arline's counsel, George Rahdert, speaking to the AIDS effect of the decision concluded that the

> Court's decision bars firing AIDS victims from most positions, because AIDS can only be transmitted through blood or semen. On the other hand, a few jobs may present a danger of transmitting AIDS. . . . Telephone linemen, for example, work in pairs and often suffer small cuts to their hands, which might expose a partner to blood through shared tools. When an AIDS victim has such a job, . . . the worker should be transferred to a position that does not present the same dangers.[41]

Nan D. Hunter, an attorney with the American Civil Liberties Union, acknowledged her reservation regarding the scope of *Arline's* application to those individuals who continue to be fired from their employment because they test positive for exposure to AIDS. "Whether they are protected," she said, "is still an enormous 'gray area.'"[42] The *Arline* decision is, of course, a construction of a questioned provision of a federal law. As such, it does not cover private employers who receive no type of federal funding. Most states have, however, enacted legislation that makes it illegal to discriminate against victims of AIDS. Interestingly, the decision also fails to address non-job-related controversies, such as school children suffering from AIDS.[43] For Ms. Hunter, the most significant aspect of *Arline* was what she termed "the court's repudiation of the Justice Department's 'fear-of contagion argument'" which she said was "inspired by the desire to permit discrimination against people most affected by AIDS, especially gay men."[44] Only time will reveal the full impact of *Arline* on the issue of discrimination of AIDs victims in the workplace.

Restrictions on Insurance Coverage

Fearing widespread efforts designed to defraud them, several of the largest insurance companies in the United States have initiated drastic curtailments in the amount of life insurance they will offer to those refusing to take blood tests for the AIDS virus. General underwriting policies deny insurance coverage altogether for those applicants testing positive for the virus that causes acquired immune deficiency syndrome. Insurance executives expressed their belief that other top companies will follow through with similar policies because they will be acting on the assumption that people with a high risk of contracting AIDS are buying policies that were double the average amounts. Although lacking conclusive proof of this practice among AIDS policy holders or applicants, a senior executive at the New York Life Insurance Company stated that "A positive test signifies a high probability of contracting AIDS." Federal scientists estimate that 20% to 30% of those carrying the Human Immunodeficient Virus, or HIV, will develop AIDS within 5 years of their infections.[45]

In a survey conducted by *Fortune* magazine and the

Allstate Insurance Company of 623 randomly selected executives, it was found that the executives were of the opinion that insurance companies, not the federal governnment, should bear the brunt of the financial burden of caring for AIDS patients.[46] That burden could be astronomical in years to come owing to the fact that in 1988 alone, about 40% of all AIDS patients will have their care paid for by federal and state funds through Medicaid at a cost of some $6 million. By 1992, these costs will soar to an estimated $2.5 billion.[47]

EDUCATION

The AIDS disease is forcing a reexamination of "the idea that homosexuality is an equally valid and legitimate life-style."[48] A member of the Mormon church's governing First Presidency told conferees to the 157th Annual General Conference that "sexual adventurism" has contributed to the spread of AIDS and, more specifically, that the "Prophets of God have repeatedly taught through the ages that practices of homosexual relations, fornication and adultery are grievous sins. Sexual relations outside the bonds of marriage are forbidden by the Lord."[49]

The fight among individuals who regard AIDS as a moral disease[50] and those who view it as a medical disease has impeded the national efforts led by C. Everett Koop, the Surgeon General of the United States, to disseminate information through all forms of media designed to educate the public regarding the disease and the methods designed to afford the opportunity to engage in "safe sex."[51] Widespread sale or free distribution of condoms is attacked as either a false solution to a critical moral problem (i.e., promiscuity) or a trivialization of sex and sexual relationships. Chastity, some suggest, should be the dominant theme in sexual ethics or family life curricula in the schools.[52] Indeed, the Federal Department of Health and Human Services "AIDS Information/Education Plan to Prevent and Control AIDS in the United States," issued in March 1987, stresses abstinence and monogamy and admonishes, "Saying no to sex and drugs *can* virtually eliminate the risks of AIDS."[53] Yet, when these policies are not followed, the Report recommends "safe sex: with the use of condoms.[54]

Massive education is, indeed, needed. Interestingly, a survey of 860 Massachusetts teenagers aged 16 to 19 found a vast depth of ignorance concerning the AIDS virus; for, many of the teens expressed their beliefs that AIDS was transmitted by kissing, sharing eating utensils or glasses, sitting on toilet seats, or donating blood. Of this survey group, only 15% had changed their sexual behavior and, of those, only 20% now follow safe sex practices of using a condom or abstaining altogether.[55]

One has to but project this Massachusetts survey in the other 49 states in order to obtain an idea of the massive scale of education needed here and the reality that, even with it, little long-lasting change in sexual behavior will be effected. It is foolhardy to pretend that propaganda and prophylactics will combat the AIDS epidemic until a vaccine or antidote is discovered. Decisive action must be taken now.[56]

The sex "taboo" is primarily due to the reluctance or failure of parents to talk about sex with their children. Because of perhaps their own limited understanding in this area, this parental attitude or perception in turn is passed on to their children.[57] All too often the unsophisticated, uneducated, and sex-fearful parents object to the basic elements of sex education even being discussed and taught in classrooms. Surgeon General Koop has urged that sexual education be taught in the schools as early as kindergarten.[58] Accordingly, "if sexuality is taught gently and gradually from an early age, it is part of your life and it doesn't come as a shock."[59] Taught a healthy and respectful attitude about sex, a greater appreciation for the fact that *diseases* are fought, not people, will be established.[60] "More and more education" is the best tool to fight ignorance and discrimination, suggested the Surgeon General.[61] The educational process is, however, a long and drawn-out affair. In 1988, the Department of Health and Human Services mailed AIDS brochures to all American households that discussed not only how the virus is transmitted, but also the importance of abstinence from sex before marriage and faithfulness thereafter as well as the use of condoms.[62]

Compulsory testing and/or screening simply must be undertaken. We must cast aside our fears of "being called anti-gay or homophobic or intolerant or puritanical or sexually unenlightened or conservative or Falwell-ish or lacking in Christian charity. Better that millions should be infected and die than that we should suffer the anguish of having such dreadful adjectives hurled at us?"[63]

AIDS-Free Health Cards: A New Gimmick

In July 1987, the British press carried the news that the entrepreneurial spirit had risen to the occasion to assist in combating AIDS and making sex safe.[64] The Weymouth Clinic in London will market cards, and provide support facilities for AIDS virus testing, with the cards showing a photograph of the holder, personal details or birth and, the all-important declaration that: "This is to certify that the bearer was declared HIV antibody-free under the recommended requirements of the World Health Organization on the dates indicated below."[65]

The Americans, as is nearly always the case, showed their capitalistic spirit sooner than the British; for in May 1987, a new report covered how the Care Card business was already in full operation.[66] An advertisement announced, "You can help prevent the spread of AIDS. Ask to see Care Card because you care about yourself. Show Care Card because you care about someone else."[67] The major difficulty with use of Care Cards and those utilized in Britain is that it can take several months for the AIDS antibodies to be picked up by a blood test. Accordingly, one could be exposed to AIDS one week, join a club and get a negative test the next, and then proceed to pass the virus on for weeks—under the illusion of "safety." So it is, then, that the same chance of tragedy exists with cards of this nature as without their use altogether.[68]

COMPULSORY SCREENING

Already, Americans by the hundreds of thousands are required by the federal government to submit themselves to AIDS testing. Indeed, by the end of 1987, the military had screened successfully 3 million members of the armed services. The Department of State's Foreign Service personnel, together with employees at the Labor Department and some 60,000 young Jobs Corps applicants are already being tested. Interestingly, several states appear to be moving forward toward mandatory testing and some proposing to withhold marriage licenses if a prospective spouse tests positive for the AIDS antibody.[69] Public sentiment appears to favor a compulsory testing program for those in high-risk groups.[70]

The Federal Center for Disease Control in Atlanta has consistently opposed mandatory testing programs—this in spite of Reagan Administration calls for such actions.[71] Instead, the Center has recommended greater use of voluntary testing for certain groups such as intravenous drug users and people considering marriage.[72] Surgeon General Koop has cautioned that mandatory testing would not only be expensive but ineffective because such a program would drive the disease underground and, furthermore, subject carriers of the AIDS virus to discrimination that could well jeopardize "housing, jobs and friendships."[73] Dr. Koop cautioned for the need to "ensure confidentiality" and urged privacy should be breached "only in the most unusual circumstances."[74] As to the issue of premarital screening, the Dean of the School of Public Health at the University of Michigan expressed her opinion that the rationale for such screening was the same "as looking for a lost object under a lamppost—because it is so easy. The people whose behavior puts them at highest risk are among the least marriage-prone group I can think of."[75]

Commercial Sex Restrictions

In an effort to curtail illicit sexual activities in commercial sex establishments—bath houses, XXX-rated movie theaters, sex clubs, and other establishments operating for the specific purpose of fostering, promoting, harboring, and encouraging multiple sexual contacts between gay males, the City Attorney of San Francisco and the Public Health Director sought injunctive relief, that was subsequently granted, designed to require such commercial sex establishments to employ monitors to survey the premises every 10 minutes in order to ascertain whether patrons were engaging in high-risk sexual activity (involving an exchange of fluids). Furthermore, proprietors were not only required to expel such offending patrons, but remove the lower portions of doors from "private rooms" in their establishments so as to facilitate accurate monitoring.[76]

Other Initiatives

On September 21, 1987, the Governor of Illinois signed into legislation a package of AIDS-related legislation[77]—together with the Sexually Transmissible Disease Control Act[78]—that, among other items, permits a "contact tracing" program which encourages individuals with sexually transmitted diseases to notify those who may have been exposed to the disease and allows such good-faith actions to immunize the reporter from civil and criminal liability. The Control Act provides specifically for strict confidentiality measures, including exemption or sexually transmissible disease information from the "Freedom of Information Act."[79] The legislation allows for the quarantine of AIDS carriers, without court authority, when individuals so endanger the public and "clear and convincing evidence" of this threat to the public's welfare is significant.[80] The AIDS Confidentiality Act requires written, informed consent prior to testing individuals for the Human Immunodeficiency Virus (HIV) and establishes confidentiality requirements for disclosure of results.[81] Further legislation requires premarital testing for AIDS be undertaken before a marriage license can be obtained.[82]

Variations and Permutations

In 1968 Dr. Linus Pauling, a Nobel Laureate, suggested that genetic carriers of sickle-cell anemia, and other recessive genes, be tattooed on their forehead or outer ear lobe so that they could be duly identified. It was hoped that such a process of identification would discourage carriers of the same defective gene "from falling in love with another" and presumably from procreating.[83]

In a March 18, 1986 opinion-editorial by the noted author, William F. Buckley, that appeared in *The New York Times*, it was urged that "everyone infected with AIDS should be tattooed in the upper forearm, to protect common needle users, and on the buttocks, to prevent the victimization of the homosexuals." He also advocated that a special AIDS blood test be conducted as a prerequisite to obtaining a marriage license and that if an intended spouse is found to carry the virus, the only way such a marriage could be validated would be for the at-risk partner to be sterilized.[84]

I myself would go one step farther and forbid the marriage altogether simply because of the public health risk of contamination for the noninfected marriage partner and because such officially sanctioned relationships would be against public policy. I would also seek to quarantine the carrier. Some carriers are more responsible than others, and I am certainly not arguing for mass detention centers or wings reserved for AIDS patients in public hospitals or prisons.[85] But, who knows what the future will hold. When one reads on the front page of the July 1, 1987 edition of *The International Herald Tribune*, that a young Los Angeles, California AIDS carrier has been arrested and charged with attempted murder for endeavoring to sell his AIDS-infected blood, second thoughts are in order about the validity and need for quarantine facilities.[86]

During World War II hostilities with Japan, the internment of second-generation Americans of Japanese ancestry (NISEI) for purposes of national security was validated by the United States Supreme Court.[87] Surely, it could be argued that the threat of destruction of the nation's health would validate a similar health internment or quarantine of AIDS carriers.[88]

What of the prostitute who uses a condom only occasionally, if at all? Should his or her civil liberties be restrained from such "professional" activities? Of course, for every right of citizenship, there is a coordinate responsibility to exercise the right *reasonably*. A recent survey of female prostitutes in the United States revealed that the AIDS infection rate was as high as 57% in some places with a national average of 11%.[89] Should the high-risk prostitute be forced to retire or otherwise be restrained in quarantine? Is sex education effective in cases of this nature?

The World Health Organization reported in July 1987 that 118 countries had reported 53,121 known cases of AIDS. This represented an increase of 1370 cases over a previous month and an increase from 111 countries in which the disease has been reported in June 1987.[90]

The first worldwide governmental meeting on AIDS was held in January 1988 and was sponsored by the World Health Organization and the British government. Although varying reporting procedures differ from country to country, it is estimated by WHO that the true number of cases worldwide is about 150,000. It was estimated that during 1988 new AIDS cases will equal this total.[91] By 1991, worldwide AIDS cases are expected to rise to 1 million, with a gradual permeation of the virus to the heterosexual population.[92] Alarmingly, by 1991, in some developed countries, the number of deaths from AIDS of men aged between 25 and 34 may well be greater than the total number of deaths in this age group from the four current leading causes of death: road accidents, suicide, heart ailments, and cancer.[93]

According to computer models prepared at the Los Alamos National Laboratories in the United States, by 1994 one American adult in ten could be an AIDS carrier. One scientist at the Laboratory went so far as to state that "AIDS represents a far bigger and more important security threat to this nation than nuclear weapons do, in any form, at this time."[94] William F. Buckley, Jr., suggests that the cost of care could rise to a $10 trillion dollar amount that, added to our present overhead, would guarantee a state of economic chaos.[95]

Given this alarming state of affairs, it was only reasonable for the Reagan Administration to call for mandatory screening or testing of couples who plan to marry, certain hospital patients, prisoners, immigrants, the military, all blood and organ donors, and those seeking treatment for drug abuse or sexually transmitted diseases.[96]

Instead of support for this bold and courageous stand, the American Medical Association rejected mandatory testing and called for new laws designed to protect the civil liberties or human rights of those who test positive for the AIDS virus and the articulation and implementation of a national policy on the issue.[97] I would respectfully suggest that the screening effort to be undertaken by the 50 states—as suggested by President

Reagan—is in fact the articulation of a national policy designed to contain the further spread of this epidemic.

It is estimated that approximately $147,000 is expended for hospital care of each patient with AIDS, and this figure does not include home medical care, outpatient care, outpatient medication, and laboratory testing or psychiatric counseling.[98]

If the predictions do in fact become a reality and by 1994 in America one of every ten citizens is infected with AIDS, how—financially—could the government cope? Using the time-honored principle of triage, should only nursing care be provided—since the factor of salvageability does not exist for AIDS patients? Or, would it be more humane and economically sound for a more aggressive policy aimed at the condonation of active voluntary euthanasia for the dying carriers be implemented? Or, would the nation run the risk of losing its soul (assuming nations have souls) if it administered euthanasia to those suffering from the AIDS virus regardless of the wishes of the carrier?[99]

The consequence of testing positive as a carrier of AIDS means discrimination and certain censure in the professional and social environments.[100] The United States Constitutional guarantee of due process and equal protection under the law together with the penumbra right of privacy are, to be sure, cherished fundamental rights of citizenship. Yet, these rights and their execution must be always balanced against the compelling interests of the state to intervene and compromise those individual rights for the maintenance and advancement of the public health and safety of the greater public.[101]

I personally have little understanding of the pleas of the civil libertarians who assert that the state has no basis for action. What is a more significant interest than promoting the public good—here, the health of its citizens—by containing a fatal epidemic? Some fear that mandatory testing is but the wedge to total abridgement of individual human rights. Or, stated otherwise, that this one excursion on the slippery slope will destroy *all* personal liberties. I see it as government action of the noblest design—action that seeks to promote the greatest good for the greatest number of citizens. Should, in individual cases, the government act excessively or beyond the bounds of reasonableness, it remains for the judiciary to correct the imbalance and return matters to normalcy.

The Kirby Legislative Commandments

Justice Michael D. Kirby, President of The Court of Appeal, Supreme Court of New South Wales, Australia, has suggested five commandments be utilized by lawmakers in dealing with AIDS: (1) to remember that local legislative and administrative responses will be ineffective unless supported by an international perspective and international cooperation and action; (2) mobilize the law to promote the prevention of the further spread of the AIDS virus (e.g., allow candid media advertisement regarding the disease, use of condoms, and the exchange of used needles and syringes by drug users for clean ones at pharmacies, make condoms readily available to all parts of society, and especially within the prisons); (3) enact laws that are cost-effective and efficient in combatting AIDS, thereby avoiding mandatory screening, detention, and quarantine; (4) design laws that avoid or minimize discrimination against those exposed to the virus; and (5) given the ineffectiveness of criminal law in prohibiting alcoholic consumption by minors, controlling pornography, prostitution, and drug use, realize that great care must be taken to avoid the overreach of the criminal law, as well as the law of torts, in dealing with the AIDS pandemic.[102] Education designed to prevent the further spread of infection, continued research designed to find a cure or preventative vaccine and a reevaluation of legal and social attitudes heretofore clouded in dishonesty, prejudice, and discrimination are the implementing policies for applying the Commandments.[103]

CONCLUSIONS

Admirable and indeed practical though Mr. Justice Kirby's "Legislative Commandments" for dealing with AIDS may be, the fatal flaw is that they miscalculate two very important matters: the ability of the international community to design and implement a cohesive strategy to combat the spread of the AIDS virus and the effect of education in containing the epidemic. If no internationally enduring plan has been found and implemented to contain violence and aggression (in Afghanistan, Lebanon, Northern Ireland, Mozambique, Nicaragua, Iran, etc.) and correct the erosion of the transnational physical environment, how can one honestly expect a level of sustained cooperation to eradicate the AIDS virus? Customs in parts of Africa that promote incestuous relationships are, for example, a stumbling block to any international effort for sustained action in combating AIDS, and such local traditions would augur ill for acceptance of the use of condoms.

The only real solution to containing the AIDS epidemic for the moment, absent the development of a vaccine of some type or other to combat it, is to identify, through screening, various high-risk groups (as targeted by the Reagan Administration) *and* physically detaining (e.g., placing in quarantine) those carriers of the virus determined to be unstable emotionally and considered potential hazards (e.g., prostitutes) to spread the virus indiscriminately to other members of society. The abridgement of civil rights must be sanctioned because of the nature of the threat to the greater world community. As observed previously, for every inalienable or fundamental *right*, there is a coordinate *responsibility* to exercise it reasonably. When individuals show that they have acted unreasonably, the government is totally justified in conditioning its response to the principle of taking action to achieve the greatest good for the greatest number of its citizens. Quite simply, then, screening and quarantine are the two most realistic avenues for present action.

Chapter 8

Noble Death, Rational Suicide, or Self-Determination

The first performance of Brian Clark's play, *Whose Life Is It Anyway*, was transmitted on March 12, 1972, by Granada television in England. It was performed subsequently on the London stage in 1978 and in New York in 1979 and made a motion picture in 1981.

Ken Harrison, a young sculptor, is the principal character of the play. As a consequence of an automobile accident, he is brought into a hospital suffering from a fractured left tibia, right tibia, and fibula, as well as a fractured pelvis and four fractured ribs—one of which punctured his lung—and a dislocated fourth vertebra that ruptured his spinal cord. While the broken bones and ruptured tissue have healed, the severed spinal cord will never be restored and thus Mr. Harrison will remain immobile from his neck down. He also suffers from depression.[1]

The play focuses on Harrison's rational choice to end his life, either in the hospital with its relative comforts, or outside its environs in his home, and the difficulties he encounters with the medical community and the law in effectuating this exercise of self-determination and at the same time establishing that his choice is rational and not the result of depression or irrational thought. In the final moments of the play, after a judicial hearing and a determination of Mr. Harrison's competency to make this choice, the attending physician allows him to stay in the hospital and promises that all treatment will stop with death resulting in no more than 6 days.[2]

Harrison's solicitor, Philip Hill, sets the central focus of this essay when he remarks:

> Perhaps we ought to make suicide respectable again. Whenever anyone kills himself there's a whole legal rigmarole to go through—investigations, inquests and so on—and it all seems designed to find someone or something to blame. Can you ever recall a coroner saying something like: "We've heard all the evidence of how John Smith was facing literally insuperable odds and he made a courageous decision. I record a verdict of a noble death"?[3]

The quest for a noble death is one that all of us must pursue during our life. Regrettably, the courts are becoming involved in this area that has traditionally been reserved to the at-risk individual, his physicians, and his family.

In June 1980, public television in the United States presented a controversial documentary on the actual case of Ms. Jo Roman, a New York City artist and writer, who undertook—with family support—an act of rational suicide in order to avoid protracted suffering from cancer.[4] Then, in early March 1983, the noted author,

Arthur Koestler, who over the years had become a strong supporter of "auto euthanasia," ended his life by ingesting a lethal dose of drugs.[5] Suffering from old age, Parkinson's disease, and other related illnesses, the 77-year-old Koestler decided that he could no longer endure. His apparently healthy 56-year-old wife also committed suicide with him.[6]

Betty Rollin's gripping biography of her mother's rational suicide, in which Betty and her husband assisted by procuring the necessary drugs for administration of the act, is a moving and indeed heartbreaking book to read. Published in 1985, *Last Wish* documents the 2½-year saga of Ida Silverman Rollin and her battle to conquer ovarian cancer. A few lines of Ida's conversations with her daughter, Betty, are significant. She said:

> "I'm not afraid to die but I am afraid of this illness, what it's doing to me. . . . There's never any relief from it . . . nothing but nausea and this pain. . . . Who does it benefit if I die slowly? . . . I'm stuck—stuck in life. I don't want to be here anymore. I don't see why I can't get out. . . ."[7]

Betty Rollin's search for accurate information to learn of drugs and their proper dosage to secure her mother's certain death without complication led her to speak with a physician in Amsterdam, Holland. In sharing with her the vital information she sought, he opines, "Yes. Modern medicine has done a great job of prolonging life, but the legal system hasn't caught up with the difficulties that inevitably arise when you have people living longer than they want to live. . . . People should have the right to end their lives when they want to, and if they need help to do it, so be it."[8]

The intense drama of Brian Clark's stage play was recreated and relived in an actual courtroom in California. For, on February 8, 1984, Judge John H. Hews of the Superior Court of California for Riverside County determined that a 26-year-old competent woman, disabled with cerebral palsy, with no functional use of her limbs except for some limited use of one hand, was not entitled to starve herself in a hospital.[9] While recognizing Elizabeth Bouvia's competency, sincerity, and rationality, the fact that she was not terminal and could live anywhere from 15 to 20 years was of controlling significance. This fact had to be balanced against her right of privacy together with the feelings of other members of society who would be offended by her act of self-determination. Particular recognition was given to the physicians, nurses, and other patients at the Riverside General Hospital where Ms. Bouvia was confined, as well as other physically handicapped persons throughout the nation. The Judge acknowledged that, "She does have the right to terminate her existence but not while she is non-terminal with the assistance of society."[10]

Two years later, on February 21, 1986, Judge Warren Deering of the Los Angeles County Superior Court ruled that Ms. Bouvia's efforts to obtain an injunction to prohibit the High Desert Hospital, a county facility where she was then a patient, from maintaining a nasogastric tube through her nostril for the purposes of increasing her nutritional intake were improper.[11] In denying Ms. Bouvia's petition, the Court ruled that the hospital's determination of the need for the use of the nasogastric tube was made in the exercise of reasonable medical judgment to avoid plaintiff being placed in a life-threatening situation."[12]

On appeal to the California Court of Appeals, the Court held that Ms. Bouvia had a fundamental right to refuse medical treatment and be force-fed.[13] Stating that it found Ms. Bouvia's condition "irreversible," there being no cure for her palsy or arthritis and that she faced 15 to 20 years of painful existence made endurable only with the constant administration of morphine, the court concluded that to maintain her as she presently was in the hospital would abridge her right of privacy, remove her freedom of choice, and invade her right to self-determination.[14]

The Bouvia Court of Appeals likened its judicial task of decision making with the case of Bartling v. Superior Court of California which was factually similar in that it involved the central issue of whether a patient could ever refuse life-continuing treatment.[15] There it was determined that the lower trial court was incorrect in its conclusion that so long as there was some potential for restoring Mr. Bartling to a "cognitive, sapient life," it was improper to allow him to refuse a proper course of medical treatment. The lower court held, specifically, that the right to have life-support equipment disconnected was limited to comatose, terminally ill patients or representatives acting on their behalf.[16]

In rendering its reversal of the lower Bartling Court *after* Mr. Bartling died, the Court of Appeals for the Second District held that a competent adult with a serious illness (here, emphysema, chronic respiratory failure, arteriosclerosis, an abdominal aneurysm, a malignant tumor of the lung, suffering from alchoholism and chronic/acute anxiety depression) had the right to order the hospital to withdraw life supports. This action was warranted even though the patient, Mr. Bartling, was not diagnosed as terminal, but was, however, regarded as "probably incurable" though with a possible potential to live a year *if* weaned from the respirator.[17] With its holding, the Bartling Court has aligned itself with the sophisticated majority of courts that uphold a patient's

right to refuse medical treatment even at the risk to his
health or his very life.[18]

The real tragedy of the Bartling case is that it was
forced upon the courts for ultimate decision making—
for Mr. Bartling had done everything conceivable to follow legal structures that would assure the maintenance
of his right of self-determination. He had executed a
living will and a separate declaration that he did not
want artificial life supports. He had executed a durable
power of attorney in his wife directing the withdrawal
of artificial supports and a release was given by the Bartling family to the Glendale Adventist Hospital and its
physicians from civil liability. The hospital argued the
case had to be presented to a court of law because it
was not only unethical for it to disconnect life supports,
but that it was subject to criminal prosecution for murder if it acted according to Mr. Bartling's wishes and
those of his family.[19] Regarding the issue of criminal
liability for their actions, the court stated that, based on
the authority of Barber v. Superior Court of California[20]
and the Florida case of Staz v. Perlmutter,[21] a hospital
no longer should have concern for its actions when such
were based on similar circumstances and directions as
those set out in Barber and Bartling.

WHEN IS "BEST" WORST?

There is an inherent paradox within any analysis of
death control. For by attempting blindly to preserve the
mere *quantity* of one's life, we may in fact degrade the
very *quality* of his life, his dying, and his final death.
But, in appropriate cases, by allowing one to die (or
indirectly hastening his death by use of a painkiller that
has the perhaps undesired although expected effect of

hastening death), we may *augment* the quality of his life, his dying, and his death—though we admittedly sacrifice some *measure* of *quantitative* living.[22]

The controlling principles for decision making in health issues should be the "best interests" principle. Stated otherwise, the physician should always endeavor to both protect and maximize his patients' best interests.[23] Inextricably related to this principle is an economic evaluation of the costs (economic, social, psychological) versus the benefits of pursuing or not pursuing a modality of treatment when major medical treatment is mandated or a response to a catastrophic or terminal illness must be made.[24] This consideration obviously expands the focus of interest to include the patient's immediate family or, in other words, those responsible for maintaining him during the rehabilitative process; it would also admit of a societal interest regarding the effective utilization of scarce or limited medical resources to competing patient needs. Here I of course refer to the doctrine of triage.[25]

Sadly, all too often, in issues of death management and control, the "best interests" principle is no longer considered efficacious.[26] Supported by professional associations fearful of incurring blanket criminal liability for their members as a consequence of outdated and unresponsive laws prohibiting euthanasia and uncompromisingly defining murder, physicians often follow procedures *not* in the best interests of their patients.[27] Added to this professional inhibition, and indeed miring it, are obtuse philosophical, jurisprudential, and ethical ramblings regarding the differences between ordinary and extraordinary means of sustaining life, intention of foreseeability of medical acts, and distinctions between acts and omissions or between killing and letting die. Reasonable actions tied to basic degrees of common sense are either obfuscated by these concerns or totally precluded from use.

Dr. Helga Kuhse of the Center for Human Bioethics at Monash University in Melbourne, an eloquent spokesman and often lone dissenter to the lock-step ideology of traditionalism, has cited a glaring example of a case where letting die (or passive euthanasia) runs counter to the best interests of a patient.

> A woman is dying of terminal cancer of the throat. She is no longer able to take food and fluids by mouth and is suffering considerable distress. She would be able to live for a few more weeks if medical feeding by way of a nasogastric tube were continued. However, the woman does not want the extra two or three weeks of life because life has become a burden which she no longer wishes to bear. She asks the doctor to help her die. The doctor agrees to discontinue medical feeding, removes the nasogastric tube, and the woman dies a few days later.[28]

Obviously, here, a painless, noble, or dignified (to that degree possible) *death* was in the patient's best interests. Yet, the *method* of bringing it about was not. The patient could have died less painfully (and, indeed, mercifully) had she been given a lethal injection.[29]

On March 15, 1986, the Council on Ethical and Judicial Affairs of the American Medical Association articulated a new policy whereby a physician can ethically withdraw "all means of life prolonging medical treatment," including food and water, from a patient in an irreversible coma. This policy goes far beyond previous AMA determinations in two major respects: (1) it specifically mentions nutrition and hydration—food and water—as a form of artificial life support, and furthermore, (2) it applies not only to terminally ill patients but to those in an irreversible coma as well.[30] Although nothing in the policy actually encourages physicians to withhold food and water, the Chairman of the AMA Council acknowledged that, "there are times, even outside terminal illness when physicians can ethically withdraw life-supporting measures, including food and water."[31] Although some critics suggest that withdrawing nutritional support from a dying or comatose patient is dangerously close to murder, the AMA statement declares that "the physician should not intentionally cause death."[32]

The President of the American Association of Neurological Surgeons observed that withdrawing extraordinary technological support from patients who have no hope of regaining consciousness can be the most humane course of treatment to follow. He continued by stating that "After a while—maybe weeks or months of seeing the patient with no concept of the present, no memory of the past and no hope for the future—a lot of families say, 'Why does this have to go on?' What's the purpose?'"[33]

I suggest that this AMA policy is what traditionally has been recognized as passive euthanasia. In order to be promotive of wider acceptance, I suggest the better position would be to consider these actions of withdrawal as but complements to basic notions of autonomy or self-determination.

Present American criminal law complicates the whole area of the non-treatment of critically or terminally ill persons. Although actual prosecutions of physicians for terminating life support in hopeless cases are quite rare,[34] still, the very possibility of criminal prosecution has a chilling effect on a physician's actions which might otherwise be promotive of patient autonomy and dignity. It can be hoped earnestly that the new

AMA policies on withdrawal of medical treatment will
have a salutary effect on law reform in this area.

Absent extraordinary circumstances (e.g., if a physician knew a medical condition of his patient was cur-
able) the disconnection of life support systems by physicians acting at the behest of families should not be
considered a criminal act.[35] Disconnection is not homicide for the very same reason that a request for cessation
of treatment by a competent patient is not suicide. Rejecting the notion that a patient's refusal of lifesaving
treatment was suicide, the Supreme Judicial Court of
Massachusetts stated:

> (1) in refusing treatment the patient may not have the
> specific intent to die and (2) even if he did, to the extent
> that the cause of death was from natural causes, the pa-
> tient did not set the death producing agent in motion
> with the intent of causing his own death.[36]

LEGAL LIMBO

The judicial system is ill qualified and, indeed, has
no special competence to consider or to make decisions
regarding treatment or nontreatment of critically and
terminally ill individuals.[37] A constant resort to the
courts in this area of concern is perhaps the counterpart
to medicine's tendency to overtreat.[38]

The family unit has traditionally exhibited the greatest degree of concern regarding the welfare of ailing
family members. The competent patient is aided by his
family when critical health decisions must be made. In
fact, family members often act as advocates for those
members confined to a hospital.[39] It is the family unit,
again, that makes decisions for its incompetent members.[40] Absent a showing of improper motives or a judg-

ment that would have a significant deleterious effect on the incompetent, the courts will not impose their evaluation of what is in the patient's best interests upon the family. And this is as it should be.[41]

Resort to the courts should only be available as a *last* resort, and only under very limited circumstances; or, more specifically, when irreconcilable conflict among family members arises, there is unambiguous evidence of wrongful motives, or a strong basis for believing that the modality of treatment or nontreatment is beyond the standard of reasonableness.[42] A hospital ethics committee or ethical tribunal should be utilized when irreconcilable differences of opinion result between the attending physician and the patient's family—this is an additional safeguard to the maintenance of self-determination without judicial interference.[43]

Law reform activity has been significant with the passage of right-to-die or living-will legislation. Some states have enacted living-will legislation that, in essence, means that in terms of overall United States population, two out of every three citizens have access to a recognized means of refusing to maintain futile prolongation of dying if they become terminally ill with no hope of recovery.[44]

A so-called "living will" is an instrument that indicates its maker's preference not to be started or maintained on a course of extraordinary treatment (sometimes specific modalities of treatment are designated) in the event of accidental or debilitating illness.[45] The biggest uncertainty surrounding living wills and their subsequent administration is related to whether health care providers are required—under pain of civil or criminal sanctions—to execute the terms of the will.[46] An interlocking concern is whether those participants charged with fulfilling the terms of the will will be assured of immunity from civil or criminal prosectuion.[47] Whether a refusal of life-sustaining therapies would constitute a suicide remains as yet another vexatious and unresolved issue.[48]

In an effort to correct some of the weaknesses and uncertainties of living will legislation, more and more states are enacting Natural Death Acts.[49] Designed to establish formally the requirements for directives to physicians in life-threatening situations, considerable difference exists among the various legislative programs regarding the assessment of penalties for either diso-beying the directive of a properly executed instrument or preventing the transfer of a patient seeking to come within the provisions of the law to a physician who will respect and follow the patient's wishes.[50]

Twenty-one states have, since 1976, enacted Nat-ural Death Acts[51] and thereby sought simply to give full legal effect to written declarations[52] by individuals who state their desire not to have life prolonged artificially if death is imminent and they are unable to speak for themselves. Thus, durable powers of attorney are rec-ognized and these state laws allow validly executed dec-larations to relieve the attending physician staff and hos-pital from civil and criminal liability for removing or withholding life-sustaining treatment[53] from an incom-petent patient in a terminal condition.[54]

Additional support for legislative reform was seen in August 1985, when the National Conference of Com-missioners on Uniform State Laws promulgated a model statute entitled the Uniform Rights of the Terminally Ill Act, that strengthens living will enactments by author-izing an adult person to control decisions regarding ad-ministration of life-sustaining treatment by executing a declaration instructing a physician to withhold or with-draw life-sustaining treatment in the event the person is in a terminal condition and is unable to participate in medical treatment decisions.

Wider acceptance by the states of The Uniform Determination of Death Act would be another strong effort to decrease if not eliminate judicial interference within this area of death management simply because of the definite posture that the legislation takes and thereby provides a framework for subsequent decision making. Under the Act,

> An individual who has sustained other (1) irreversible cessation of circulatory and respiratory functions or (2) irreversible cessation of all functions of the entire brain, including the brain stem is dead. A determination of death must be made in accordance with accepted medical standards.[55]

A Bold Initiative

The Hemlock Society of California is presently campaigning to place in 1990 as a referendum initiative an amendment to the State Constitution that would recognize and expand "the inalienable right of privacy" to include "the right of the terminally ill to voluntary, humane, and dignified doctor-assisted aid in dying."[56] The subsequent act emerging from this proposed amendment is termed the Humane and Dignified Death Act and structures a mechanism that would allow adults to

> . . . execute a directive directing the withholding or withdrawal of life-sustaining procedures or, if suffering from a terminal condition, administering aid in dying. The directive shall be signed by the declarant in the presence of two witnesses not related to the declarant by blood or marriage and who would not be entitled to any portion of the estate of the declarant upon his death under any will of the declarant or codicil thereto then

existing or, at the time of the directive, by operation of
law then existing. In addition, a witness to a directive
shall not be the attending physician, an employee of the
attending physician or a health care facility in which the
declarant is a patient, nor any person who, at the time
of the execution of the directive, has a claim against any
portion of the estate of the declarant upon his death.[57]

For declarants who are patients in a "skilled nursing
facility" to make a valid directive, one of the two witnesses must in fact be a patient advocate or ombudsman
designated as such by the California State Department
of Aging.[58] Provision is made for the revocation of the
directive by the declarant "without regard to his mental
state or competency"[59] and an exculpatory clause prohibits an imposition of "criminal or administrative liability on the part of any person for failure to act upon
a revocation . . . unless that person has actual knowledge of the revocation."[60]

No civil, criminal, or administrative liability is imposed upon a physician or licensed health professional
acting under the direction of a physician who fails "to
effectuate the directive of the qualified patient . . . unless he wilfully fails to transfer the patient to a physician
who will comply with the directive."[61] More specifically,
actions of withholding "or withdrawal of life-sustaining
procedures from or administering aid in dying to a qualified patient," under this Act will not be regarded as
acts of suicide.[62] Provision is made for amending the
Civil Code, Section 2443, by providing that actions
under the Humane and Dignified Death Act shall not
be construed as condoning, authorizing, or approving
mercy killing or permitting "any affirmative or deliberate act or omission to end life other than the withholding or withdrawal of health care pursuant to a durable power of attorney for health care so as to permit

the natural process of dying."[63] The proposed amendment to Section 2443 of the Civil Code makes it clear
that "an attempted suicide by the principal shall not be
construed to indicate a desire of the principal that health
care treatment be restricted or inhibited or that aid in
dying be given." It should be stressed that the Proposed
Act does not condone, authorize, or approve mercy killing or permit any affirmative or deliberate act or omission to end life *other than* by a licensed physician and
when requested by the patient pursuant to a properly
executed legal document.[64]

The Model Directive proposed by The Hemlock
Society to accompany this proposed legislation in Article
Two states:

> In the absence of my ability to give directions regarding
> the termination of my life, it is my intention that this
> directive shall be honored by my family, agent and phy-
> sician(s) as the final expression of my legal right to 1)
> REFUSE MEDICAL OR SURGICAL TREATMENT and
> 2) TO CHOOSE TO DIE IN A HUMANE AND DIG-
> NIFIED MANNER.

Provision is made in Article Three that if a diagnosis of
pregnancy is made and known by the physician, the
directive will have neither force or effect during the
course of the pregnancy.

Article Four stipulates the declarant's understanding that he has a terminal illness for which it is not likely
he will live with treatment *more than 6 months.* In Article
Five, the declarant's agent is given "full power and authority to make health care decisions" to the same extent
as the declarant would if capacitated. In exercising this
authority, the agent is required to make those decisions
consistent with the declarant's desires as specified in the
directive "or otherwise made known" to the agent "including, but not limited to" the desires of the declarant

"concerning obtaining or refusing or withdrawing life-prolonging care, treatment, services and procedures, and administration of aid in dying."

The Humane and Dignified Death Initiative is an eloquent effort to build upon the Living Will legislation, Natural Death Acts, and durable power of attorney acts to finally recognize—in a definitive legislative scheme—that the continuation of life for terminally ill persons under conditions of severe pain and suffering constitutes not only severe cruelty and disregard for human dignity but an invasion into basic rights of privacy and self-determination. It can be but hoped that the Humane and Dignified Death Initiative sponsored by The Hemlock Society will in fact be accepted by the citizens of California and, as such, begin to structure a framework for principled national decision making in this central area of human values and ethical concerns.[65]

All of these various legislative actions make clear that society is at last beginning to recognize the rights of its citizens to control their ultimate destiny in a reasonable manner. Rational self-determination should thus be regarded as an inalienable right.

CONCLUSION

I find the very use of the word "suicide" abhorrent, for the historical cultural perspective associated with this act precludes an understanding or acceptance of the reasons for the act being committed. An act of suicide defies reason—so the traditional view goes. It is, quite simply, always an irrational act. Contemporary thinking discards this view. I suggest that the taxonomy of suicide be discarded and, in its place, acts of self-destruction, within the context of this essay, be viewed simply as acts of self-determination.

What a person is, what he wants, the very determination of his life plan, and his concept of goodness, fairness, or equity are truly the very intimate expressions of autonomy or self-determination. And, by asserting one's responsibility for the results of this self-determination, substance is given to the fundamental concept of liberty.[66] Indeed, as John Stuart Mill wrote so eloquently, the only reason for interfering with one's "liberty of action" is "self-protection," for his absolute right of independence extends "over himself, his body and his mind."[67]

If we as concerned scholars, teachers, and health care providers can begin to accept this revision of our thinking and in turn proselytize the message inherent in this reclassification of suicide, we can set a new tempo for social, cultural, legal, and medical discourse and action. Thus, the concept of suicide, and particularly rational suicide, would give way to a wider acceptance of self-determination and those associated acts taken to ensure noble death.

Chapter 9

Procreational Autonomy

Values Gone Awry?

A number of commentators have observed that the liberalization of state adoption laws, coupled with the liberation of women,[1] or the fundamental constitutional rights of privacy and procreation may demand or indeed establish the right of unmarried women to be artifically inseminated[2] or of a single woman to participate in the process of surrogation and become a surrogate mother.[3]

The approach taken here is to present an argumentative analysis that shows that there is no "fundamental right" to artificial insemination of unmarried women, and that the statutes limiting this practice to married women are reasonable and reflective of sound public policy considerations. Statutes that might, furthermore, seek to legitimize the status of surrogate mothers could undermine those fundamental policies that recognize the family as the very essence of societal strength and well-being. My purpose, then, is not to explore the complex legal and ethical interstices of the new reproductive biology.[4] Rather, one aspect of this process is to be examined; namely, the process of artificial insemination of unmarried women either for their own personal purposes of unfettered pregnancy or under the guise of surrogation for an infertile woman. Surrogation thus becomes an analytic complement to the so-called sexual privacy of some women as they seek to express their sexual freedom[5] or procreational autonomy with the aid of unconventional (e.g., artificial) means to become pregnant, and must be evaluated as well.

ARTIFICIAL INSEMINATION: OVERVIEW

There are three kinds of artificial insemination (AI): (1) artificial insemination with the husband's sperm or homologous artificial insemination (AIH),[6] (2) artificial insemination by a donor or heterologous artificial insemination (AID),[7] and (3) artificial insemination by the combination of the donor's and the husband's sperm (AIC).[8] Although the reasons that lead to this form of reproduction may vary from case to case, AI remains the simple medical procedure of inserting a syringe into the vagina and injecting semen toward the opening of the uterus.[9]

Traditionally considered an alternative to adoption, permissible only within the marriage relation,[10] AI has shed its social shackles as unmarried heterosexual and lesbian women have sought to become AI mothers. As a result, one commentator has recently observed: "The use of A.I. by unmarried women for the first time allows reproduction to be separated from sexual activity and traditional family life, forces the law, itself rooted in traditional values, to face the clash between 'the power of science over human reproduction and traditional family values.'"[11] The judicial response to this question has been somewhat limited, but by no means indifferent, while the legislatures of 22 states have limited the practice of AI to married women.[12]

The Judicial Response

Only one reported case, C.M. v. C.C., concerned the artificial insemination of an unmarried woman.[13] This case, truly a judicial encounter of the first kind,[14] presented a most unusual set of facts. C.C., an unmarried woman, desired to have a child but did not want to engage in premarital sex. She considered artificial conception with friends, but C.M., whom she had been dating for two years, suggested that she use his sperm since they had often spoke of an eventual marriage. The couple visited a local sperm bank, but were denied the use of its facilities. While consulting with the doctor, however, C.C. familiarized herself with the procedures utilized for AI, namely, a glass syringe and a jar. For the next several months, C.C. made regular visits to C.M.'s apartment for this purpose. C.M. would stay in one room, obtain sperm, and then give it to C.C., who would artifically inseminate herself in another room. Conception was eventually achieved, and sometime during the third month of her pregnancy, this psuedo-Platonic relationship ended.

When the child was born, C.M. brought an action for visitation rights to the child, asserting that he was the natural father and expressing his desire to fulfill that role. Consequently, the court had to decide whether C.M. was the natural father of the child "or whether he should be considered not to be such because the sperm used to conceive was transferred by other than natural means."[15]

The court refused to take a position regarding the propriety of the use of artificial insemination between unmarried persons.[16] Nonetheless, it recognized that public policy considerations dictated that, whenever possible, it was in the child's best interests to have two parents. Accordingly, C.M.'s petition was granted, with an attendant responsibility for the support and maintenance of the child.

On the question of an unmarried woman's right to AI, this case cannot be said to provide a favorable precedent. Although purporting to take no position on the subject, the court in arriving at its conclusion implicitly repudiated the condonation of this practice by an unmarried woman. The court could have justified its decision on a number of legal theories recognizing the rights and efforts of C.M. in the insemination process such as implied contract, quasi-contract, or estoppel.[17] Given the magnitude of the interest at stake, however, the court properly shied away from basing its decision on any property or contract theory, and unequivocally recognized that the "contraband stork"[18] was no match for the nuclear family. That is to say, a *child's interests* in having a father and mother were paramount, no matter how it was conceived.

Legislative Responses

With the growing number of children being born each year as a result of AID,[19] the need to clarify their legal status has become a growing concern of many state legislatures. Some states have enacted statutes[20] defining the rights of children born through this procedure.[21]

All of the statutes, except that of Oregon, limit the practice of AID to married women. Oregon's statute arguably authorizes this practice by implication for unmarried women, since it provides, in part: "Artificial insemination shall not be performed upon a woman without her prior written request and consent of her husband."[22] Nonetheless, when the statute is viewed as a whole, and given that it has yet to receive any judicial construction, its legislative meaning remains uncertain.[23]

The equal protection clause of the Fourteenth Amendment prohibits the states from denying to any person the equal protection of the law. Since a state statute represents state action for equal protection analysis,[24] such a statute must conform to constitutional strictures that limit AID to married women or otherwise violate the equal protection clause. As subsequent analyses will demonstrate, these statutes are not amenable to constitutional attack.

SURROGATION

Simply defined, a surrogate mother is a woman (single or married) who conceives artificially from a married man and carries a pregnancy for that man's wife because of her infertility, which is a barrier to the marital fulfillment of having children. Upon delivery, the surrogate relinquishes all of her control over the child and gives it over to the contracting couple, normally for adoption.[25] Thus it is seen that maternal surrogation is, in reality, a logical counterpart to artificial (donor) insemination of AID.[26] The surrogate mother, as the artificial father/donor, participates in no physical act of sexual intercourse. In those cases where a wife gives her consent to a surrogate mother arrangement, as where a husband would consent to his wife's participation in donor insemination (AID), there is no basis for considering the fertilization techniques used in both procedures to effect a pregnancy as adulterous.[27]

None of the states have passed laws specifically on the issue of surrogate motherhood regulation. Yet a number of them make it a crime to pay anything of value to a parent in consideration for obtaining consent to adopt or obtain custody of a child and impose a heavy penalty or imprisonment for violation thereof.[28] In this way, some deterrent exists to prevent extensive "black market" operations in adoptions.[29] Since the legality of a contract is tested by and depends upon the place where it is in fact made,[30] a jurisdiction having such a law would hold a contract entered into between a husband and wife with a surrogate mother to relinquish her parental rights to an infant born of such a contract for subsequent adoption by the natural father and putative mother to be illegal and thus invalid. Thus, the bargain is viewed in terms of an *adoption* contract. If this same contractual relation is viewed as a contract *to bear* a child, less objection and greater acceptance of the contract itself should be recognized simply because the purchaser is the natural father of the child. With such a consideration, fears of commercialization become less worrisome as a competing or undermining factor over the interests of the child or the biological mother.[31] Interestingly, an unmarried man would encounter even less legal entanglement in dealing with a surrogate. Since no wife would be involved, it would be unnecessary for a man to adopt his own child. He would simply pay the surrogate without risk and, as natural father of the child, take custody of it upon birth without any formal adoption procedure being required.[32]

Presumptions of Paternity

The Uniform Parentage Act presents an initial obstacle to the establishment of paternity in surrogate contract situations. Section Five of the Act controls the use of artificial insemination and provides: "The donor of semen provided to a licensed physician for use in artificial insemination of a married woman other than the donor's wife is treated in law as if he were not the natural father of a child thereby conceived."[33] While it is obvious that this provision was crafted in order to protect anonymous AID donors from all legal responsibility for those children fathered as a consequence of their donations of semen, if the provision is adopted in toto, this statutory language could well establish a difficulty for the real father under a surrogate contract to establish either paternity or to assert parental rights of visitation, for example.

Children born of a validated marital union are presumed to be the legitimate issue of that union. All of the states have, regardless of their adoption of the Uniform Parentage Act, recognized this presumption.[34] Thus, if a surrogate were married, the issue she bears will be presumed—unless rebutted by proof beyond a reasonable doubt—to be the legitimate child of herself and her husband and not of the artificial donor of the sperm. The donor or biological father could bring a custody suit and upon proof (again, beyond a reasonable doubt) that he was in fact the biological father, it would then fall to the court to determine which parent or set of parents could better serve the long-term interests of the child.[35]

Additional Policy Issues

As has been seen, there is an obvious if not parallel relationship between AID and contracts to bear children. Yet, while donor identity is generally assured of confidentiality in heterologous insemination, the identity of a surrogate mother is generally known by the contract couple. Appearance, in fact, is often a major consideration—together with the prospective surrogate's medical history, education level, environment, and cultural background—in making a judgment as to the suitability of a surrogate.[36] Because of the increased potential for genetic anomalies in births by women over 35 years of age, this age is usually the cutoff for surrogate candidacy. The vailability of the pool candidates is furthermore limited normally to women who are presently married or who have been divorced.[37] Single, unwed women should be regarded as not suitable simply because their involvement would not only be regarded as promoting immorality (and perhaps, technically, adultery) but, as a practical matter, an unproven record of birth successes might promote a further element of uncertainty for the contracting parties, which is undesirable. The right of a single, unmarried woman to control her own physical autonomy is, for some, an ever-evolving concept measured in proportion to the degree of state interest in preserving the public welfare and morals. The right of a married woman or the extent to which she can act or conduct her marital affairs, *without* the informed consent of her husband—or, specifically to submit herself to surrogate mother status—remains an open-ended legal question.

Perhaps an even more unsettling issue is the extent of the surrogate's autonomy during the period of the pregnancy versus the extent and nature of the right of control of her and the fetus by the biological husband and his wife. If, for example, after agreeing to abstain from uses of alcoholic beverages during the pregnancy, the surrogate does in fact imbibe on a regular basis, could a court orer be obtained to stop such consumption?[38] If so, how could it be enforced? By total restraint (i.e., hospital confinement)? Suppose the surrogate did not reveal her propensity to consume alcohol (or prescribed drugs) and the child is born with a genetic impairment or defect which is determined to be a direct consequence of the actions of the surrogate. Could the surrogate be sued for negligence? Suppose, further, neither the husband nor his wife wish to take the defective child as their own and the surrogate does not want it? Should a penalty be assessed against all concerned parties because of this "misdeed"? In such a situation, the infant becomes a ward of the state and thus a responsibility of the taxpayers if and until an adoption can be arranged.

If a physician were negligent in failing to screen adequately a prospective surrogate mother candidate and the infant is born with a genetic deficiency, could a suit for malpractice be obtained against the attending physician or the surrogate herself?[39] Or, suppose a surrogate decided to keep the contract baby. Could she in turn sue the biological father for child support?[40] As a practical matter, it would appear that in a case where a court decreed the biological surrogate mother had a right to keep "her" child and, assuming the mother had insufficient funds to support herself and the child, that financial support by the biological father would be in the best interests of the child and of the state.[41]

Emerging Case Law

The informal opinion in 1978 of Wayne County Juvenile Court Judge James Lincoln, relative to the legality of surrogate mother status, held that a volunteer could bear a child for a couple to adopt, but the state law forbade any payment of fees to the surrogate for her service.[42]

In the Circuit Court of Wayne County, Michigan, an opinion of considerable import and interest was rendered by Judge Roman S. Gribbs on January 28, 1980.[43] Jane Doe, her husand, John, and May Roe, a would-be surrogate mother, sought a summary judgment which, if granted, would have allowed them to execute their agreement to have Mary conceive a child with John, through artificial insemination duly administered by a physician, and subsequently upon birth, allow the Does to adopt the child. In consideration for her services, Mary was to receive the sum of $5,000 in medical expenses. At issue was the constitutionality of two Michigan statutes which made it a criminal offense to "offer, give, or receive any money or other consideration or things of value in connection with . . . placing a child for adoption."[44] The plaintiffs, in addition to seeking a declaration of unconstitutionality as to the two statutes, sought to enjoin the defendant State Attorney General from prosecuting them for proceeding with their agreement with Mary Roe.

Plaintiffs' challenge of the statutes on the ground of unconstitutionality was tied to an argument that urged the court to void the statute because of its vagueness and, furthermore, because of the statutes' invasion of their consitutional right of privacy. The court concluded that for a statute to be violative of due process, it must prescribe conduct which is so vague that an average person of normal intelligence would be forced to guess at the meaning of the statute itself. The statute in question was determined to be sufficiently direct in order to give fair notice to all those affected by it.[45]

The second argument in the plaintiffs' case is that the statute in question not only invades their constitutionally guaranteed right of privacy, but fails to comply with basic requirements of compelling state interest and the strict drafting required of statutes which regulate an act within the privacy right. Acknowledging that the only "fundamental" rights within the Roe v. Wade privacy doctrine may be so considered, the court concludes that the specific activities of marriage—which include, procreation, contraception, family relationships, child rearing, and education—are not the bases of plaintiff's action. Rather, plaintiffs maintain a selective attack of the statutes which prohibit only the exchange of money or other valuable consideration in the adoptive process. The protection provisions of the statute used to effect a legal adoption are not challenged. This forces the Court to conclude that a contract which utilized a statutory grant of authority to effect a child's adoption and provides for valuable compensation in connection therewith, is not within the blanket protection of the right or privacy.[46]

The court then proceeds, albeit through *dicta*, to discuss more fully the right of privacy issue.[47] Noting that intrusions into the right of privacy may be tolerated when compelling state interests so merit and when such intrusions are drawn narrowly in order to articulate the specific state interests at stake, the Court observes that even if such a fundamental right of privacy were applicable here, the state's actions to prevent commercialism in child adoption matters is both proper and an overriding legitimate state interest.[48] "Baby bartering" is totally abhorrent to the public policy of the state of Michigan. Here, specifically, the money which the plaintiffs wish to pay the surrogate was intended as an inducement for her to conceive a child not normally intended to be conceived and, as such, would be a violation of the state's public policy against such acts. Thus the Court held against the plaintiffs and observed that any change in the present area of contention would have to be initiated by the legislature.[49]

On November 9, 1980, a baby boy was born in the Louisville, Kentucky, area and subsequently became the first recorded baby born to a surrogate mother *under contract*.[50] The following April, the infant was, by Jefferson County Court order, legally adopted by his putative mother and his biological father. While identities of the new parents were not disclosed, the facts show that an Illinois housewife, whose pseudonym is Elizabeth Kane, was hired for money to carry the issue of her artificial insemination from a married man's semen. After waiting 90 days after birth, as required by Kentucky law, the putative mother filed for adoption of the infant. Her attorney declared that the adoption order was final and a lawsuit maintained by the State Attorney General of Kentucky challenging the legality of the surrogate contract would have no effect.[51] Presumably, the State's Attorney is maintaining his action on a statutory provision that prohibits advertising or soliciting children for adoption and accepting any remuneration for procuring any such child for adoptive purposes. No penalty is specified, however, for violation of this provision.[52]

On February 3, 1988, the New Jersey Supreme

Court determined in the matter of Baby M that a $10,000 surrogacy contract between William Stern and Mary Beth Whitehead was for the sale of a child or, in the alternative, of a mother's right to her child, and was thus invalid and unenforceable.[53] The original trial court had found the contract legally enforceable and the surrogacy relationship protected constitutionally unless it could be shown such was not in the best interests of the baby born from the contract and, furthermore, allowed the wife of the genetic father to adopt the questioned infant immediately.[54] Here, the State Supreme Court allowed that such contracts could be recognized as nonetheless legal, however, so long as the relationship involved no fee and allowed the surrogate to reconsider or change her mind about relinquishing all rights to the baby in question. Although allowing the trial court's grant of custody to the natural father, the High Court remanded the case for a determination of the rights of visitation of the natural mother[55] and urged the legislature to deal "with this most sensitive issue" and recognized the marked difficulties in legislating on the subject that, while private, was of "such public interest." A final solution to the conundrum can, the Court acknowledged, be settled only "when society decides what its values and objectives are in this troubling, yet promising area."[56]

EQUAL PROTECTION FOR WHOM?

When a statute structures a classification which results in the disparate treatment of similarly situated people, the equal protection clause demands that the courts inquire into the justification for the classification. The level of scrutiny to be employed, however, will be contingent on the interest at issue. In the traditional case, the legislative classification will be upheld if it is "reasonable, not arbitrary,"[57] and if it rests "upon some ground of difference having a fair and substantial relation to the object of the legislation."[58] If a fundamental right is involved, however, a more critical examination of the classification is required.[59] The level of scrutiny here demands that "the statutory classification . . . be not merely rationally *related* to a valid public purpose but *necessary* to the achievement of a *compelling* state interest."[60]

The AID statutes make an express classification between married and unmarried women. Accordingly, the next step in the analysis requires a determination of whether the right affected by the classification is fundamental. If the right at issue is anywhere "explicitly or implicitly guaranteed by the Constitution,"[61] it will be deemed fundamental, and the strict scrutiny standard will apply. Such will also be the case if the classification involves a suspect class.[62]

Equal protection clause arguments are faulty as a basis for those seeking a means to protect informal, illicit relationships.[63] Since the Eisenstadt case, in fact, classifications based upon marital statutes have been upheld unanimously.[64] The legal discriminations allowed between the married and the unmarried are justified by the fact that such acts and procedures of discrimination determine both the ability of the state to contain and regulate marriage as a social institution, as well as maintain and validate those various legitimate individual interests which are inherent in every marital relationship and include property interests, taxation, contracts, and torts.[65]

A number of commentators have, however, recently suggested that an unmarried woman's right to

AID may be discerned from a liberal reading of the Supreme Court decisions establishing fundamental rights to procreation and privacy.[66] It is to an analysis of those cases that discussion turns.

As society evolves and changes, so too do many of its values.[67] Autonomy, self-representation, personhood, identity, intimacy, and dignity are all essentials of privacy.[68] The extent to which these essentials play a role in shaping a degree of sexual, procreational autonomy must surely remain largely flexible, for to attempt to define them with precision would challenge and erode any efficacy they might enjoy.[69] The right of the state to control and to shape the behavior of both individuals and of groups regarding the birth of children is always an area of high and legitimate concern.

The major view is that private conduct between consenting adults or, for that matter, personal conduct of any nature, should be regulated only to the extent necessary to prevent harm to others.[70] Conformity is thus not a value of momentous concern and certainly not a value worth pursuing.[71] The opposing view is that the business of the law is suppression of vice and immorality simply because if violations of the very moral structure are indulged and promoted, such actions would surely undermine the whole basis of society itself.[72] Under the former view, the state would be arguably justified in acting to control personal decision making, if not for the need to prevent illegitimates from proliferating, then to prevent the ultimate economic harm to society of having to help bear the expenses associated with the maintenance and education of a fatherless child born of artificial insemination. Similarly, the prevention of harm theory could be invoked in surrogation where the state, by preventing such acts, seeks to maintain the dignity and continuity of the family unit.[73]

A Basic Right to Procreate?

The first case to address tangentially what has now come to be regarded as a fundamental right to procreate was Buck v. Bell.[74] In Bell, the Supreme Court upheld a Virginia statute that permitted the sterilization of inmates in state institutions who suffered from a hereditary form of insanity or imbecility.[75] This opinion, authored by Justice Holmes, was written before the development of the fundamental right/compelling state interest standard.[76] Thus, it must be determined whether the court's opinion implicitly recognized the existence of a compelling state interest, or whether the Court merely failed to perceive procreation as a fundamental right.[77] The latter appears to be the case; indeed it has been suggested that the Court's pervasive emphasis on the state's right to promote the general good or welfare approximates a rational basis standard of judicial review.[78]

In Skinner v. Oklahoma[79] the Supreme Court again had occasion to consider the validity of compulsory sterilization laws. Unlike the Court in Bell, which found no equal protection[80] violation, the Skinner Court struck down Oklahoma's Habitual Criminal Sterilization Act on equal protection grounds. The statute provided for the sterilization of habitual criminals, i.e., anyone convicted of three felonies, but did not consider felonies which arose from the violation of the prohibitory laws, revenue acts, embezzlement, or political offenses.[81] The Court initially recognized that marriage and procreation are fundamental to both the existence and survival of mankind.[82] It then went on to observe, however, that a classification distinguishing larcenists from embezzlers, for purposes of criminal sterilization, represented a form of invidious discrimination. Consequently, the Court subjected the classification to strict scrutiny and found it violative of the equal protection clause.

Although a number of Supreme Court decisions[83] have since cited the Skinner case as at least validating, if not in fact creating, a constitutional right to procreate,[84] it is important to recognize precisely the contours of that right. In both Bell and Skinner, the Court was confronted with sterilization statutes. Sterilization, unlike other methods of control over human reproduction, is irreversible.[85] Thus, in discussing the procreative "right" affected by Oklahoma's Habitual Criminal Sterilization Act, the Skinner Court aptly observed that this "right [was] basic to the perpetuation of the race."[86] Given this background, it is seen that the procreative right recognized in Skinner was simply a right to remain fertile, and not an uninhibited right to engage in potentially procreative conduct. Subsequent decisions which have focused on a fundamental right to privacy have further delineated the contours of this right.

Searching for a Fundamental Right to Privacy

The Constitution nowhere mentions a right to privacy. Nor is any right of sexual freedom to be found within the gambit of procreative rights recognized by the Supreme Court; nor for that matter has the Court fashioned a general right of personal privacy which is sufficiently broad-based to permit sex outside marriage.[87] In Griswold v. Connecticut,[88] however, the Supreme Court for the first time recognized a constitutionally protected zone of privacy, and invalidated part of a Connecticut statute forbidding the use of contraceptives by married persons.[89] The protection of this aspect of procreative autonomy "was largely subsumed within a broad right to marital privacy"[90] which "stressed the unity and independence of the married couple and forbade undue inquiry into conjugal acts."[91] From this, however, it cannot be argued that there must exist a corresponding fundamental right to reproduce or to the use of artificial reproductive technology.[92] As Justice Goldberg made emphatically clear in his concurring opinion, Griswold "in no way interfere[d] with a State's power of regulation of sexual promiscuity or misconduct," and thus the constitutionality of Connecticut's statutes prohibiting adultery and fornication remained beyond dispute.[93]

In Einsenstadt v. Baird,[94] the Court was confronted with a Massachusetts statute that prohibited the distribution of contraceptives to unmarried persons. In holding that the statute violated the equal protection clause of the Fourteenth Amendment, the Court observed that "If the right to privacy means anything, it is the right of the *individual,* married or single, to be free from unwarranted governmental instrusion into matters so fundamentally affecting a person as the decision whether to bear or beget a child."[95] Accordingly, the Court in Eisenstadt fleshed out the procreative skeleton of Griswold which initially appeared confined to the so-called "sacred" precincts of the matrimonial bedroom chambers.[96] This decision, however, did no more than to refine a qualified right to procreative autonomy blurred by the Griswold Court's emphasis on the marital relation.[97]

In Roe v. Wade,[98] the Court squarely addressed an integral part of the individual's right to procreative autonomy when an unmarried woman in a class action suit challenged the constitutionality of the Texas criminal abortion laws. The Court articulated a new source of privacy derived from the Fourteenth Amendment's standard of personal liberty and inherent restrictions upon state action and held that this right was sufficiently broad to embrace a decision made by a woman whether or not to terminate her pregnancy.[99] The court went on to state, however, that it was not recognizing "an unlimited right to do with one's body as one pleases."[100]

The final pertinent case of interest in this area is Carey v. Population Servs. Int'l.[101] In Carey, the Court invalidated a New York Statute which regulated the sale and distribution of contraceptives to minors and stated that "at the very heart of [the] cluster of constitutionally protected choices," recognized in the previous privacy cases,[102] was "the decision whether or not to beget or bear a child."[103] This decision is particularly instructive on the question of the unmarried woman's right to artificial insemination, for it examines the previous privacy cases and delineates the extent of the individual's right to procreative autonomy.

It has been suggested by some commentators that since a woman has a right to terminate her pregnancy and to use contraceptives, a posteriori, the conduct required to bring about those procreative choices must also be protected.[104] The Court's opinion in Carey indicates, however, that this is simply not the case.

First, with regard to contraception and abortion, the Court made clear that it is "[the] individual's right to decide to *prevent conception* or *terminate pregnancy*" that is protected.[105] Such unequivocal language, however, lends little or no support to the argument that a concomitant right to conceive is also protected. Second, the Court emphasized that its decision did not encompass any constitutional questions raised by state statutes regulating either sexual freedom or adult sexual relations.[106] This reading of Carey is supported by a later decision of the Court which stated that if "the right to procreate means anything at all, it must imply some right to enter the only relationship in which the [s]tate . . . allows sexual relations to legally take place."[107] The lesson from the Court' decisions in Skinner, Griswold, Eisenstadt, Roe, and Carey is plain: "procreative autonomy includes both the right to remain fertile and the right to avoid conception,"[108] but nothing more.

The Level of Scrutiny and the State's Justification for Action

Since the unmarried woman's decision to be artificially inseminated does not fall within the gambit of any recognized fundamental right, the state statutes limiting this procreative technology to married women "[may] be sustained under the less demanding test of rationality. . . ."[109] Under this test all that is required is that the distinction drawn be "rationally related" to a "constitutionally permissible" objective.[110] In employing this rather relaxed standard, courts must be sensitive to the fact "that the drawing lines that create distinctions is peculiarly a legislative task and an unavoidable one."[111]

Absent a suspect classification or the infringement of a fundamental right, the Supreme Court has recognized that legislation "protecting legitimate family relationships" as well as both the regulation and protection of the family unit are "venerable" concerns of the state.[112] Thus, statutes limiting the availability of artificial insemination to married women fall squarely within this classification.

As early as 1888, the Court recognized marriage as "the foundation of the family and society, without which there would be neither civilization nor progress."[113] Recently, the Court observed that "a decision to marry and raise a child in the traditional family setting must receive . . . protection."[114] Thus, although certain aspects of an individual's right to procreative autonomy have correctly been divorced from the familial and marriage relationship, the Court has also implicitly recognized that, whenever possible, childbearing should take place within the traditional family unit.[115] An unmarried woman's decision to seek artificial insemination goes against the tide of these pronouncements.

An instructive analogy may be made to the law of adoption. As the statutes regulating artificial insemination, adoption statutes have their genesis in state law.[116] Although all states currently allow adoption by unmarried adults,[117] it occurs only in rare cases. In the adoption of H., an unmarried[118] middle-aged woman sought to adopt a 13-month-old child, for whom parental care by a young couple was available. In rejecting her application, the court observed:

> Adoption by a single person has generally and in this Court's experience been sought and approved only in exceptional circumstances, and in particular for the hard-to-place child for whom no desirable parental couple is available. In the universal view of both experts and laymen, while one parent may be better than none for the hard-to-place child, joint responsibility by a father and mother contributes to the child's physical, financial, and psychic security as well as his emotional growth. This view is more than a matter of present convention, anthropologists pointing out that the institution of marriage, which is a method of signifying commitment to

> such joint responsibility, evolved in response to the need
> for two-parent care of children.[119]

This observation applies clearly with equal force to the
case of artificial insemination for an unmarried
woman.[120] Indeed, if a state may reasonably regulate
unmarried adults in their quest to adopt children, it
would be anomalous to suggest that it could not regulate
the use of a procreative technology designed to bring
children into the world.

More importantly, however, the unmarried woman's access to artificial insemination and, thus, surrogation, directly undermines [t]he basic foundation of the
family in our society, the marriage relationship. . . ."[121]
The desirability of having a child reared within a traditional family unit has been repeatedly recognized by
the courts. Moreover, it is clear that "[w]ithin the traditional model, marriage serves as the genesis of the
family. . . ."[122] Accordingly, the inherent procreative
potential of this union,[123] together with the stability that
this provides to the social fabric[124] would be dealt a fatal
blow by permitting unmarried women to be artificially
inseminated or to act as surrogate mothers.[125] Equally
unpersuasive is the argument that a state is painting
with too broad a brush when it limits AI to married
couples. Although the Supreme Court has failed to for-
mulate a concrete definition of the family, Moore v. City
of East Cleveland[126] represents the only clear extension
of protection" routinely afforded to the nuclear family
"to a quasi-familial group."[127] In Moore, a zoning ordinance which limited an area to single-family dwellings
was challenged by a woman who shared her home with
her two grandsons. The Court merely recognized that
the extended family occupies a place in American tradition similar to that of the nuclear family, and, thus,
is to be guaranteed protection by the Constitution.[128]

As the procreation and privacy cases illustrate by analogy, however, the fact that a mother and her offspring may find protection within the nuclear family structure does not imply a right freely to bring about that condition, nor does it demonstrate that the limitations placed on AI with respect to unmarried women are in any way irrational. Thus it assuredly demands an expanded definition of family in order to contend that statutes limiting AI to married women are not rationally related to a constitutionally permissible objective. The line of demarcation may be drawn imprecisely, but the Constitution is not offended "simply because the classification 'is not made with mathematical nicety or because in practice it results in some inequality.'"[129]

CONCLUSIONS

The legal system, by protecting relationships as kinship and formal marriage, promotes not only those interests of the parties involved, but the interests of society in those social and political structures which ensure a long-term individual view of liberty.[130]

In judicial decisions affording familial and marital relationships a higher degree of constitutional protection, traditions have played a pivotal role. In the procreative field, the Supreme Court has carved out a limited degree of autonomy for the individual. As demonstrated, a woman's fundamental right to privacy or procreation does not encompass a right to AI or to surrogation. Accordingly, statutes limiting the use of this reproductive technology need only be rationally related to the promotion of a constitutionally permissible state interest. A state's desire to raise children in the *traditional* family setting, while at the same time promoting the institution of marriage and the family, is an unquestionably permissible if not laudable objective. In 1952 Justice Frankfurter cautioned: "Children have a very special place in life which the law should reflect. Legal theories and their phrasing in cases readily lead to fallacious reasoning if uncritically transferred."[131] The legislatures, in limiting the practice of artificial insemination to married women, have taken this counsel to heart. The extended use and application of this procedure through surrogation must be strictly controlled by legislative design. Surrogation should only be tolerated by a married woman, with her husband's actual consent, and then only under proper medically supervised standards. As a medical aid to infertility, surrogation should then only be allowed as an adjunct to medical treatment of this impediment, and not as a popular or novel experiment.

A legislative program designed to validate, and thereby license, the procedure of surrogation for married women as well as the married surrogates participating therein would not only seek to protect the health and well-being of the issue born but assure the safety of the surrogate herself. Such a legislative program would ideally include provisions shaping the rights and determining the extent of the liabilities of the contracting parents in the surrogate compact vis-à-vis the infant. Due consideration should be given to shaping the sphere of responsibility for various types of error which intermediaries, such as doctors and lawyers, might commit in facilitating the whole process. Ideally, the specific policy matters coincident with the administration of a surrogation program, once structured, would be implemented by an administrative body or licensing board. The Surrogate Parenting Associates, Inc., of Kentucky could well serve as a model for legislative adoption throughout other states. The policies and standards for evaluation in processing requests for surrogate mothering are both comprehensive and equitable in their design and utilization.[132]

The new reproductive biological techniques for parenthood portend an enormous significance for humanity and demand the need for a searching inquiry into the parameters for future development.[133] The legislative branch of government is far better equipped to deal with this inquiry than is the executive or judicial. Thoughtful study and a cautious plan of action is needed now before advancing complexities overwhelm, confuse, and confound the role of the rule of law in meeting the challenges of the brave new world of tomorrow.

In 1983, the Federal Attorney General of Australia observed that the Hawke Government recognized the dual role of law in simultaneously reflecting and molding the social conscience of the community.[134] The importance of this statement cannot be overrated, for it presents an obvious and simple truism: namely, it is impossible to be pragmatic and objective when considering complex issues of great emotional moment such as, for example, procreational autonomy through artificial operatives.[135]

Justice Michael D. Kirby has observed wisely that a coherent institutional response to bioethical quandaries of the nature analyzed previously is needed, rather than momentary or ad hoc resolutions.[136] Instead of making the hard decisions that are called for here in the chambers of isolated courtrooms, he suggests that the open community be the forum of decision making.[137] Toward the attainment of this end, some notable successes have been charted that are designed to assay community views and thus assist in the formulation of principles and the development in this fluid area of concern.

Since 1981 three Australian national public opinion surveys conducted by the Australian affiliate of Gallup International have shown high rates of public approval of
both artificial insemination and in vitro fertilization.[138]

Enlightened lawmaking can only be enlightened if
it has a solid informational or data base upon which to
draw for a sustained level of development. In the final
analysis, then, it is to be remembered that, "It is not for
lawyers to direct the community, but for the community
to direct the legislation into such fields as they desire
the law to operate."[139]

EPILOGUE: FURTHER THOUGHTS ON
SURROGATE LIABILITY

In 1884, with the case of Dietrich v. Northampton,[140] it was held that a fetus was unable to maintain
a cause of action for in utero injuries because, as a part
of the mother, no legal duty is owing to one not yet in
being. With the case of Scott v. McPheeters[141] 1939, the
first case was recorded allowing for recovery for prenatal
injuries. This was followed in 1946 with Bonbrest v.
Kotz[142] that established viability as the central factor in
permitting recovery. The scope of recovery was extended to nonviable fetuses with the holding in Kelly
v. Gregory[143] where it was determined that when a child
is born alive it is entitled to obtain relief for any prenatal
injuries sustained by it *any time* after the point of conception owing, as such, to another's negligent conduct.[144]

Generally, as a consequence of the parent-child immunity doctrine, negligence actions brought by an unemancipated child against its parents are barred.[145] Nine
exceptions have been grafted onto this doctrine, how

ever, ranging from situations where the family relationship no longer exists and the parent tortiously injures the child in an intentional, wanton, or grossly negligent manner, to where the parent is covered by a liability insurance policy.[146]

A child's right to maintain a legal action against his surrogate mother's behavior that consequently negligently inflicted prenatal injuries upon him, should not be hindered or barred by the parent-child immunity doctrine. In this regard, it is important to recognize the fact that the surrogate mother's contractual agreement to carry the infant for the ultimate benefit and consideration of the contracting parents imposes upon her a much higher standard of care than the imposed socially and legally upon a mother fulfilling the *traditional* childbearing role. Indeed, the surrogate's behavior during her pregnancy is obviously of great concern to the contracting parents and focuses directly upon the best interests of the embryo/fetus.

Inadequate nutrition, consumption of alcohol, tobacco smoking, and drug ingestion, to list the major behavioral patterns that directly threaten the development of the fetus, must be regulated. Excessive abuses of these actions must be checked and legal liability imposed when injury is threatened or occurs.

If a court can impose a duty in prenatal injury actions to the father of an unborn infant for willfully failing to furnish necessary food, clothing, shelter, or medical attendance for his child,[147] it is but reasonable to place a surrogate mother within the same zone of responsibility. Since the surrogate mother assumes an affirmative and unabridgable duty to act reasonably and thus promote or insure the best interests of the fetus at the very moment of her impregnation, if injury results subsequently and it is shown the injury was proximately caused by a breach of that duty, the surrogate must be held liable for her negligent conduct.[148]

Although evidentiary proofs of causation are obviously going to be difficult in cases of this type of consideration, they are not insurmountable. Accordingly, while the legal cause of the injury one sustains must be traceable to the wrongful actions of another for a cause in negligence to be established, the legal cause need not be either the sole or the predominant cause of injury.[149] It is only necessary to show the offending conduct by the actor is a substantial or material factor causing the injury.[150]

While it must be conceded that, tragically, not all women who become surrogates possess the same levels of intelligence or sophistication, imposing liability is not to be premised upon an actor's lack of knowledge or understanding, but rather because she failed to acquire that knowledge is cases of this matter.[151] Thus, for a surrogate to assert that she was unaware of the consequences of smoking, drinking, or other harmful acts is indefensible; for, quite simply, knowledge is expected of an individual serving as a surrogate mother, whether or not expressly provided for in the contract or imputed to her by reason of her status. "The surrogate exhibits negligent behavior by failing to acquire the knowledge which is an essential part of her task as a surrogate mother."[152]

Chapter 10

The Case of the Orphan Embryos

The major news story of June 18, 1984 in Australia, and indeed around the world, was the discovery that two frozen embryos might well become, if successfully implanted in a surrogate mother, heirs to an estate left by the death of what was thought originally to be their biological parents.[1]

The facts revealed that, in 1981, a Los Angeles, California couple, Mario and Else Rios, participated in the in vitro fertilization program of Melbourne's Queen Victoria Medical Centre. Then 50 years old and infertile, Mr. Rios allowed a local yet anonymous donor from Melbourne to artifically inseminate three eggs from his 37-year-old wife; one being implanted and the other two frozen for future possible use. Mrs. Rios miscarried and was not emotionally stable to undertake subsequent implantations at that time. Before she could return and endeavor to use the other embryos, she and her husband died in a Chilean plane crash. Under the California laws of intestate succession, applicable because no will was executed by the Rios, Mr. Rios' son by a previous marriage is entitled to his father's share of the estate and Mrs. Rios' 65-year-old mother takes her daughter's share.[2]

The central issues raised here are: do the two frozen embryos have any legal rights to live and be implanted in a surrogate mother and, when and if born, to assert inheritance rights in the Rios' estate? Equally important is the very issue of the extent to which research into the new reproductive technologies should be allowed or be restricted.[3]

THE COMMON LAW

Early in its development, the Common Law, and especially the concept of quickening as it was articulated in the criminal laws, found inextricable relationships existing with theology.[4] While early Christian teachings stressed the sanctity of life from its beginning at fertilization, this particular view was modified subsequently and a distinction made between an *embryo formatus* and an *embryo informatus*.[5] Thus, life was regarded as commencing when an unborn infant first undertook movement in the womb, or when it quickened and was thus infused with a soul.[6] Consequently, the early Common Law scholars maintained that only after the fetus quickened could a destruction of it be classified as murder.[7]

The Australian Posture

The multidisciplinary Medical Research Ethics Committee of this National Health and Medical Research Council has reviewed divergent community views concerning both the moral and ethical status of fetuses and concluded that while the traditional legal position may be more definite in some respects,[8] it is quite vague in others.[9] Nevertheless, it was concluded that prior to birth *and* separation from its mother, a fetus (or an embryo) has but potential or contingent civil legal rights, as well as enjoying limited protection under the criminal laws of abortion and child destruction depending upon a particular stage of gestation.[10] Indeed, in an interesting judicial corollary, Mr. Chief Justice, Sir Harry Gibbs, of The Australian High Court, ruled in March 1983, "that a fetus has no right of its own until it is born and has a separate existence from its mother."[11]

Imperfect Civil Rights and Inadequate Criminal Laws

None of these imperfect, potential, or contingent civil rights or the criminal laws of abortion or child destruction would have, obviously, any application to a frozen embryo before implantation; for the law has refused to recognize the moment of fertilization as the point at which legal rights are conferred.[12]

In two states of Australia, New South Wales and The Australian Capital (A.C.T.), there is authority to suggest that no legal protections are afforded the fetus until it has developed for a period of 28 weeks.[13] Judicial opinions in other states have suggested that the fourth or fifth month of pregnancy is the pivotal point at which full legal protection will be given a fetus.[14]

The Deputy Chairman of The New South Wales Law Reform Commission and a widely respected figure in the field of the New Biology, has stated that any claims of the Melbourne "orphan" embryos against the Rios' estate are "fanciful."[15] Agreeing with Mr. Scott, the Attorney General of the State of Victoria, stated that the embryos have no legal status of any nature. Furthermore, if they were successfully implanted and were subsequently developed, they would be the legal offspring of the surrogate "mother" and the man to whom she is married.[16] Presently, absent any direction by the decedents before death, the Rios' embryos are the legal responsibility of the hospital where they are kept.[17]

THEORIES OF RECOVERY

There are four possible legal theories under which considerations could be given, hypothetically, to the Rios' embryos. First, they could be viewed as personal property and be allowed, as such, to pass by the intestate laws of succession to the heirs of the Rios' family, thereby allowing them to do with the embryos whatever they wished. The difficulty here is that for embryos to be considered personal property, they must be recognized in law as having an *economic* value.[18] Obviously, such a determination is, at this stage, impossible to make.[19]

Secondly, the embryos could be treated as though they were fully developed children and subject to the appointment of a guardian *ad litem* by a court in order to determine what would be in the "best interests" of the embryos vis-à-vis their implantation or destruction. Thirdly, since it is unlikely under existing law that the law, itself, would elevate the frozen embryos to a legal status of some type of "personhood," one could accept Mr. Chief Justice Gibb's previously stated position and regard the embryos, from a proper legal standpoint, as being non-entities.

Finally, the Queen Victoria Hospital could be recognized as the constructive trustee for the deceased Rios' and, accordingly, be allowed to decide in justice the fate of the embryos. Constructive trusts, also referred to as implied trusts, do not arise because of the expressed intent of the party (settlor) or parties (settlors) executing a formal trust. Rather, they are created by a court of equity or a court exercising equitable powers in order to prevent unfair or untoward acts inconsistent with the perceived intention of the parties in question from occurring.[20]

Existing Guidelines

For 2 years, guidelines have existed in Australia, developed and approved by the National Health and Medical Research Council, which cover the ethical problems associated with in vitro fertilization programs. Guideline Seven specifically suggests that an upper time limit be placed upon the storage of embryos which goes beyond "the need or competence of the female donor."[21] Applied, then, to a woman's capability to conceive, it is obvious that at Mrs. Rios' death, her capability has ended and the two embryos could be destroyed. Interestingly, the Council, supported by the Australian Medical Association, endorsed in 1982 not only the use of in vitro fertilization as an acceptable scientific procedure to correct infertility among married couples, but the use of donor eggs in women to produce embryos and the use of artificial insemination by anonymous male donors.[22]

The Reform Movement

Since there appears to be no legal power to prevent the fertilization of human eggs in a laboratory, private or otherwise,[23] and, given the growing realization that courtrooms are an improper forum for resolving complex philosophical dilemmas based on competing scientific and technological developments,[24] what remains for the future here on this front? Can an accommodation be reached along some humane, equitable, or objective lines?

In early 1982, law reform activity in the field of the new reproductive biology began in earnest under the vigorous leadership of Russell Scott of the New South Wales Law Reform Commission who headed an advisory committee studying artificial insemination. A few weeks later, the Victorian government established a Committee, subsequently designated the Waller Committee, chaired as such by the distinguished Sir Leo Cussen, Professor of Law at Monash University, Louis Waller, whose mandate was to investigate the problems arising from in vitro fertilization and donor gametes (the male sperm and female eggs). Soon to follow in similar research activities were the Queensland Government and that of Western Australia.[25]

The Waller Report on the Disposition of Embryos Produced by In Vitro Fertilization was released in mid-August, 1984, in Melbourne, Australia, by the Attorney General of Victoria. The Committee concluded, among other points, that: the disposition of stored embryos is not to be determined by the hospital where they are in fact stored;[26] that such embryos are not to be regarded as possessing legal rights or having rights to lay claim to inheritance;[27] and in cases where "by mischance or for any other reason, an embryo is stored which cannot be transferred as planned, and no agreed provision has been made at the time of storage . . . the embryos shall be removed from storage."[28] The Committee held additionally that: embryos could be frozen[29] and that experimental research "shall be immediate and in an approved and current project in which the embryo shall not be allowed to develop beyond the state of implantation, which is completed fourteen days after fertilization."[30] Some of the recommendations of the Committee will be incorporated in the Victorian government's in vitro legislative proposals for subsequent Parliamentary adoption, while others will be open to further debate and study.[31]

The standing committee of Attorneys-General of Australia has agreed on the desirability to work toward the development of a uniform code of legislation cover in vitro fertilization and the legal status of children born through the use of donor semen or ova.[32] One can but guess whether a working consensus will ever be reached among the six state governments allowing for a uniform code of regulation. Terms such as when life begins must be agreed upon before regulations can be written.

A REPRIEVE AND A NEW BEGINNING

Disregarding the recommendation of the Waller Committee regarding the Rios' "orphan" embryos, the legislature in the State of Victoria enacted a law which directs that an attempt be made to have the embryos implanted in a surrogate and, if subsequent birth results, the child be placed for adoption.[33] Although restricted in application to the two Rios' embryos, the plan would have obvious repercussions for the future development of policy in this field.[34] Thus, any children resulting from the embryo implants would, under this new Victorian legislation, be taken to be children only of the adoptive parents.[35] The extent, if any, to which the law of Victoria would impact upon a legal action in California would be quite speculative.

Legislative Realities

The New South Wales Parliament passed the Artificial Conception Bill which was given Royal Assent on March 5, 1984, and states that a child born from in vitro fertilization where genetic material is provided by the husband and wife, or where semen is provided by a donor, will be deemed the child of the husband and wife. The same principles are also applied to de facto relationships. Yet, interestingly, the law does not cover children born as a result of an in vitro fertilization procedure using donated ova.[36]

A partial legislative response to the confusion of the IVF procedure has been posited by the Victorian Parliament. Read March 20, 1984, before the Legislative Council, the Infertility (Medical Procedures) Bill of 1984 legalized the IVF procedure for married couples after a waiting period of anywhere from 12 to 24 months, during which time the couple is examined in order to determine whether these procedures are the only available means of achieving pregnancy, and are counseled regarding the potential chances for success and/or failure. Provision is made to allow donor sperm, if the husband is infertile, or donated ova where the wife is incapable of producing eggs. As drafted originally, the proposal carried no declarations or protections for frozen embryos.[37]

A Network of Safety

Sir Gustav Nossal has suggested that instead of cumbersome legislative restrictions on use and development of new reproductive technologies, a "network of safeguard" is, in some respects, already in place and needs to be tightened. Continuing public debate of the issue of scientific freedom versus governmental regulation is the focal point of the "network." Additional components include the further development of safety standards, self-regulating industrial guidelines and ethical codes and reliance upon the Common Law mechanism of "reasonable case" to shape and determine the extent to which experimentations and scientific investigations will be allowed. Finally, greater reliance upon national committees within the Ministry of Science and Technology for overseeing Australian research in this field is urged.[38]

A noted American authority suggested action be undertaken at three levels: the enactment of model state legislation which defines with clarity the identity of both the legal mother and the father of all children, including as such those born to other than their genetic parents; the development and promulgation by concerned professional organizations of guidelines for sound clinical practice, and the establishment of a national body of various experts in law, medicine, science, ethics, and public policy, whose mandate would be to monitor ongoing developments in the area itself and report to Congress on an annual basis regarding the desirability of specific legislative schemes and regulatory plans.[39]

CONCLUSION

The Nossal approach is sound, but it can be strengthened by legislative and regulatory programs[40] of the design and nature of the present undertaking in Victoria which seeks to validate and control IVF procedures. If a creative approach to problem resolution can be proferred here by drawing the best elements of the Nossal approach and the Law Reform efforts being pursued presently in Australia, the Continent will be able to boast a full and active partnership between law and scientific medicine—with the legacy of such an undertaking redounding to the benefit of all mankind.[41]

Chapter 11

Science, Religion, and the New Biology

COMPATIBILITIES AND CONFLICTS

Science has been defined as, "intelligence in action with no holds barred."[1] It began as the simple pursuit of truth, but today is fast becoming incompatible with veracity, quite simply because complete veracity leads to a form of complete scientific skepticism.[2] Science was originally recognized, and indeed valued, as a method to know and understand the world.[3] Ever since the time of the Arabs, "science has had but two simple functions: to enable us to know and learn about things and to thereby assist us in doing things."[4] Now, as a consequence of the development of the scientific method and the triumph of technique, since it is viewed as a means of changing the world.[5] Probabilities are at the center of scientific inquiry. As such, an absolute form of truth is not within its scope of realization. Yet, science can yield such a high degree of probability that it becomes a certainty for all practical purposes.[6]

Science is a way of ordering experience; it is ordered knowledge. Its constant testing and referral to the facts of past experiences should be viewed as the only valid way man can progressively increase both his knowledge and control of the objective world.[7] This constant reference to past experience in the quest for knowledge is the most significant attribute of the scientific method, for from it comes "the cosmic side of that intellectual scaffolding of religion we call theology."[8] Nonetheless, there has been a prolonged conflict between religion and science.[9] Perhaps one of the basic reasons for the inherent conflict has been the differences in focus of religious creed and scientific theory. A creed is said to embody both eternal and absolute truth. Scientific theory is always recognized as tentative, with modifications sooner or later found necessary. The scientific method, then, unlike the religious creed, is one which is logically incapable of arriving at an ultimate statement.[10]

Religion, to a considerble extent, consists in a way of "feeling" sometimes more than in a set of beliefs. The beliefs are secondary or supportive of these feelings.[11] There are some things people believe, then, because they feel as though they are true;[12] and such feelings and beliefs are a source of mystery and incomprehensibility to the scientific mind. Faith is an unknown and rather primitive principle to the scientist.[13] From the standpoint of maintaining its strength, efficiency, or power, religion must face change in the same spirit as science does. While religion's principle may be immutable and eternal, the expression of those principles requires a continual development.[14]

Roman Catholicism, predominantly under the leadership of the late Pope John XXIII, has charted a new course of contemporary expression, particularly in its liturgy. Certain dogmas, such as the Virgin Birth by Mary, papal infallibility, priestly celibacy, the exclusivity of the male priesthood, and the sanctity of creation, remain inviolate. The sanctity of creation, however, has presented problems to the scientific community as it explores eugenic proposals and fetal experimentation.

During the middle and latter half of the nineteenth century, science made its greatest inroads into religion. Then a credibility gap was beginning to open between what could be explained within the framework of religion and what could be explained within the scientific frame of analysis. Some view this gap as continuing to widen simply because the more scientific discoveries about the universe that are made, the less explicable they become. Around the mid-twentieth century it was generally believed that gradually science was attempting quite successfully to explain the entire universe. The more scientific facts presented for understanding, the more knowledge of the universe would emerge. In the latter part of the century however, there is a concern because rationalists and humanists are suggesting that within the near future science will not be able to say anything fundamental about the true nature of the universe.[15]

The advancement of science is often blamed for a loss of religious faith.[16] There is, contrariwise, a belief that the work of science has been the one factor causing the greatest understanding of religious truths today.[17] The overriding fact to be observed is that normally a scientific advance will show that statements of various religious beliefs, if they have contact with or are tied to physical facts, require some sort of modification either through expansion, reinterpretation, or restatement. If the particular religion is grounded in a sound expression of truth, the required modification will only "exhibit more adequately the exact point which is of importance."[18] A contradiction, in formal logic, is the signal of a defeat. In the evolution of real knowledge, a contradiction marks but the first step in progress toward a victory, and this is the principal reason why a variety of opinion is tolerated and even encouraged.[19]

The equivocal attitudes of Christians regarding their religious faith cannot be so easily modified. These attitudes are complicated by suspicion, ignorance, and misunderstanding: suspicion directed against advancing technology which appears to have a considerable power for good or evil depending on the technologist who directs it; ignorance from not knowing sufficiently the true nature of science and technology; and misunderstanding of the Christian doctrine of creation which has in turn led to false ideas about materialism.[20]

As viewed today, there is no actual conflict between the statement of theological principles and the scientific method of inquiry by investigation, because there is no interrelationship or mutual dependence.[21] Based on revelation and faith, theology presents its concepts and principles totally independent of the scientific theories about nature or speculations regarding the past.[22] Both science and religion present different phases of human activity and embody distinctive experiences. While religion is fundamentally a spiritual experience, science is based on "sensuous experience."[23] Yet, science and religion are one in the experience of revelation they offer to those who pursue them: the revelation of a supreme fact of mental or progressive spirit and experience.[24]

In the final analysis, the scientist and the theologian depend mutually on experience and interpretation. They ask different types of questions—not expecting to receive the same types of answers in return. Science and religion are but reflections of different aspects of man's social experiences. If one can move beyond popular misconceptions regarding the nature and role of science and religion, he will find no conflict between their methods of study and practice.[25] Religion should be devoted to the expression and fulfillment of final values beyond which no other values can exist.[26] A scientific approach to religion, then, becomes but a noble effort to study the true story of man, the relation to the source of his being and his duties, privileges, and structure of values. Science, if pursued within this construct, provides the basic framework for a new dynamic testament, a new scripture of truth about man and his destiny.[27]

If the administration of science is to be perfected for the betterment of mankind, not only are moral ideals needed, but a spiritual vision as well. The most notable scientific work has flowed consistent with a high conception of social duty and with a spirit of altruism. Science is but a means to an end, with its values being determined by the end.[28] Societal progress as expressed in the law must, in the ultimate analysis, embrace two complementary plans of development; plans embracing both scientific research as well as increased moral understanding and appreciation.[29]

THEOLOGICAL CONSIDERATIONS

The Roman Catholic View

Roman Catholic dogma teaches that marriage does not bring to the married couple an absolute right to children—only a conditional right. All that may be done is for the couple to avail themselves of the use of legitimate medical processes in order to assure their sexual act be performed in a natural way in order to attain "its fertile union."[30] The Church thus stresses the fact that coition be recognized solely as an act designed for procreation, between husband and wife only, and that the act itself be unimpeded by direct means. "Human sexual congress in order to be authentic, must involve intravaginal ejaculation by the husband and retention of the semen, or at least no deliberate effort at expulsion, by the wife."[31]

The exclusivity of the marriage contract forbids intercourse with a third person or the use of semen from a donor to effect artificial congress. Thus, the Church considers the use of A.I.D. (heterologous insemination by a donor) to be adulterous irrespective of the fact that a husband may consent to his wife's indulging in sexual "relations" with another man through artificial processes.[32] The major point of emphasis is the invasion by a third party into an exclusive marriage contract. The unity of love and procreation must remain inviolate.

The normal way of obtaining semen is through masturbation. But this very act is considered to be a "perversion of the sexual faculty" because it is not procreative.[33] If semen were collected from the wife's husband (AIH or homologous insemination) in a manner other than through autoerotic techniques,[34] and then injected into the wife's reproductive tract, it has been submitted that Church teaching would regard this act as valid, since love and procreation are not really separated, but indeed furthered by the act.[35] It is a physical disability which forces the husband to resort to AIH in the first instance. It is love which induces him to seek artificial means of impregnating his wife. The unity of love and procreation is thus, thereby, strengthened.[36]

In contradistinction, fertilization by donor gametes in vivo or in vitro would be automatically rejected by the Church because although no adulterous relation was present, two different communities would be created: one procreative and the other loving. Although perhaps anonymous, the donor becomes a silent partner in an exclusive relationship which admits no intruders.[37] Yet, the technological manipulation of a husband and wife's own gametes would appear to be compatible with the principle of loving and procreation, since the basic marital relationship remains intact.[38]

Although a new intellectual climate of openness and reevaluation is evident in the Roman Church hierarchy, this climate has not fostered new and significant moral directions for the Church and its theologians in this specific area of concern.[39] The official Church posture today remains the same as that first announced by Pope Pius XII in his address to the Fourth International Convention of Catholic Physicians, October, 1949. The Pope stated that an act of artificial insemination outside the state of marriage was immoral; use of a donor or third party's semen (AID) to facilitate conception by a married couple was also immoral. Such an act was to be "rejected summarily." The Pope also rejected use of AIH or homologous artificial insemination for Catholic couples.[40]

In 1951, Pope Pius XII, addressing the Congress of the Italian Catholic Union of Midwives, sought to amplify his views regarding AIH. Accordingly, he expanded upon his idea that the conjugal act was a personal act of "simultaneous and immediate cooperation on the part of the husband and wife." He continued by observing that "this is something much more than the union of two seeds, which may be brought about even artificially, without the natural action of husband and wife."[41]

The concern of Pope Pius XII over the manner of

obtaining semen in AIH is, today, no longer viewed by a number of moralists as a valid obstacle to this procedure. Indeed, when this method of conception is the only method by which the "procreative mission" may be met, pastoral counselors are encouraged to suggest use of AIH.[42]

Donor insemination or AID, however, is still regarded by many as "an intrusion into the exclusivity and intimacy of the conjugal bond that is hard to reconcile with the Christian understanding of the nature of conjugal love."[43] Nevertheless, there is clear evidence that couples who have successfully used AID have enriched their personal and marital lives and that the issue has not been "a painful reminder" of the husband's impotency.[44]

In a September 4, 1978 article, the editors of *Time* magazine noted that Albino Cardinal Luciano, before he assumed the papacy, appeared to have adopted a modern understanding of the scientific imperative in the brave new world of the coming twentieth century.[45] Although the experimentations that led to the birth of the first test tube baby were severely criticized and condemned outright by some Church theologians, Cardinal Luciani commented that if the husband and wife who participated here "acted in good faith and with good intentions, they could even gain great merit from God" for their actions.[46]

The Cardinal sought to balance this viewpoint, however, by further elaboration on the extent of the scientific mandate, noting that science must be sufficiently regulated in order to prevent an industry directed toward the manufacture of children. Acknowledging that the dictates of individual conscience must be followed in cases of this nature, he cautioned that "a well-informed conscience—does not have the duty of creating law, but of informing itself on what the law of God dictates."[47]

Before Cardinal Wojtyla's elevation to the papacy as Pope John Paul II, he too had gone on record in his book, *Love and Responsibility*, published in 1960, as against all artificial methods of birth.[48] Although known as a staunch conservative on specific issues of doctrine, morality, and church authority, in the same book, the Cardinal also recognized sexual pleasure deriving from the marital relation.[49] While breaking no new ground in Roman Catholic ethics or doctrines, the Pope recently cautioned that scientists engaging in a wide range of medical procedures such as artificial insemination and genetic engineering to be aware of "the implicit danger to the rights of man" from discoveries and advances in these fields, for such actions could well violate the individual's physical and spiritual life.[50]

The Protestant View

The conservative Protestant Ethic maintains that some acts are specifically commanded in the Bible and must be followed by all. The literalist approach to the Bible has serious weaknesses as a basis for religious ethics primarily because often the moral precepts found within the Bible are both unclear and contradictory.[51] Under conservative Protestantism, a monogamous marriage is the biblical expression of God's unalterable will. The only alternative to marriage is abstention from sexual intercourse.[52] The only inferences which may be drawn from this philosophy is that AID is morally objectionable as an invasion of a monogamous marriage unity and that genetic engineering qualifies as an offensive sexual relation.[53]

The liberal and more contemporary Protestant view is that since all of the biblical commandments are ambiguous and, thus, not clear expressions of God's will, there are no universal modes of conduct required of Christians.[54] In defining relationships between persons, the crucial determinant is whether love is present or absent. Therefore, the validity of one's actions sanctified and legalized by a marriage contract is of secondary importance. What is of central importance is whether coition is a truthful expression of a personal commitment to one another: Is it honest and carried out in such a manner so as not to exploit the other person?[55] So long as mutuality of life is expressed, then almost any procedure within the gambit of a practice of the "New Biology" would be tolerated.

Whether AID is considered adulterous is really only a question of semantics.[56] AID involves a far more responsible level of decision making than the "normal" one-night-stand act of adultery or the clandestine relationship. No infraction of the marriage vows is promoted by a consensual decision regarding the use of AID by a married couple. Moreover, when a husband allows his wife to be impregnated by a donor, it is this very consent and desire for offspring which assures that the subsequent child itself is of primary concern. There can be no allegation of broken faith in such a situation. In an adulterous relationship, the very essence of that relationship is grounded in broken faith by one partner to the marriage contract.[57] In such a situation, should a "careless mistake" be made and issue result as a consequence of the exaltation of physical and emotional needs outside the bounds of the marriage, that "mistake" is usually the subject of concern, despair, and non-acceptance instead of love and acceptance as in a consensual act of AID.

The Jewish View

Under Jewish law, a woman who participates in AID is not guilty of adultery. The child born of the artificial act is regarded as legitimate, regardless of whether its mother is married or single.[58] Only when it is established conclusively that a child has been born of an adulterous or incestuous relationship is the child regarded as illegitimate.[59] There is a strong presumption against adultery or incest. In fact, it is virtually impossible to prove any conception was adulterous or incestuous since the husband is always presumed to be the father of his wife's children.[60]

Interestingly, the donor in an act of heterologous insemination, although in no way stigmatized by his act, remains the natural father of the child and can never rid himself totally of this relationship. However, he may be relieved of liability for support of the issue and his estate removed from claims of inheritance by children whom he normally would never know or see.[61] This strict standard of civil liability obviously does not preclude the development of a foster parent relationship in addition to the natural relationship.[62]

CONCLUSIONS

What is clear in the theology of the New Biology is that belief in God and the perceptions of the Divine Will are not shared uniformly. Religious groups will disagree within their own religious ranks regarding the use and application of the New Biology. Because of this variance, perhaps it is better and wiser to submit opinions only about specific moral applications of genetic engineering rather than focus in on positive condemnations of one development and its use as opposed to another.[63]

Religions should be careful not to be too terribly dogmatic in the area of the New Biology. A religious ethic rooted in human well-being should survive the pressures of the emerging brave new world. Churches and religious teaching will either be molded to reflect the new ethics of the age or will simply die out:

> [R]eligion in the age of science cannot be sustained by the assumption of miraculous events abrogating the order of nature. Instead, we should see acts of God in events the natural causes of which we fully understand.[64]

It is time for the major religions to advance a balanced scientific spirit of inquiry, investigation, and basic reevaluation and thus provide the law with a much needed point of direction. If law and religion can but jointly approach the problems of the "New Biology" there is an excellent likelihood that a degree of stability will emerge. If accepted, the utilization of artificial insemination will assist family planning and ensure the continued "sacredness" of the family unit in those cases where, without its use, no family would be forthcoming.[65] Given this progressive attitude of enlightenment, the use of artificial insemination within the bonds of matrimony can only serve as complement to a Catholic society both of today and tomorrow. Again, by endeavoring to effectuate a balancing test which seeks to minimize human suffering and thereby maximize the social good, a eugenically sound standard of qualitative life and a continued recognition of the sanctity of creation can exist together.

EPILOGUE: ROME SPEAKS

Dated February 22, 1987, and issued March 10, 1987, the Vatican released its "Instruction on Respect for Human Life in Its Origin and on the Dignity of Procreation: Replies to Certain Questions of the Day" over the signature of the Prefect of the Congregation for the Doctrine of the Faith, Joseph Cardinal Ratzinger.[66] Abortion and experiments on fetuses were condemned as well as *all* forms of artificial fertilization (i.e., in vitro fertilization, embryo transfer) for married couples, as being unnatural interferences with the physical act of intercourse. Given the fact of previous rejections by Rome of donor artificial insemination (AID) and even husband artificial insemination (AIH), the current sweeping condemnations of all reproductive technologies and their use was not unexpected.[67] In addition to insisting on the total inviolability of germinating life at the very instance of fertilization, the document rejects absolutely surrogate mothers.[68] When asked why the sweeping denunciation of all new reproductive technologies and specifically the use of surrogate mothers, in vitro fertilization and embryo transfers *within* a marital relationship of love and affection, Cardinal Ratzinger termed their use as unloving and egoistic. A prominent Jesuit theologian termed the Cardinal's response as "utter nonsense."[69]

In order to bypass the rigidity of the Congregation's condemnation of the technology of procreation, a modified G.I.F.T. procedure has been advocated. With this method, termed gametes interfallopian transfer, an egg is removed surgically and placed beyond the fallopian blockage to allow an in vivo fertilization. But, in order to blunt the denunciation of Rome here, a perforated condom that retains semen at the tip is used during intercourse and, after the act has been concluded, the imprisoned sperm is collected by a syringe and again mechanically aspirated toward the egg.[70] This compromise is termed "repugnant" and reduces "a healthy morality to an embarrassing moralism."[71]

The Congregation's document has been termed "reactionary" in that, "It is a defensive reaction against the possibility that not all traditional articulations of the unity of sex, love and parenthood measure up to the experience of people whose lives are sexual, marital, parental and also Catholic Christian."[72] Although this pronouncement from Rome should not be regarded as definitive or final, but rather a contribution to an ongoing reflection,[73] the fact remains that a certain rigid and almost unyielding nonscientific mindset is exhibited within the pages of this "Instruction." If a sustained dialogue between Rome and its worldwide congregation—and particularly the childless North American married couples—is to be developed, it will not be memorialized by such documents as this "Instruction." Sadly, the barren Catholic married couples will not be guided by it, for the heartbreak of a childless marriage and efforts to relieve that heartbreak will be given a higher ordering than unquestioning obedience to a distant edict from Rome.

Notes

CHAPTER 1

1. G. Orwell, *1984* (Signet Classics ed. 1984).
2. V. Ferkiss, *Futurology: Promise, Performance, Prospects* 5, 46 (1977).
3. *Id.*
4. S. Mcauliffe & K. Mcauliffe, *Life for Sale* 32–34 (1981).
5. *Id.*
6. *Id.*
7. Gladwell, Washington Area Expected to Retain Leadership Role in Biotech Industry, *Wash. Post Bus. Mag.*, Mar. 7, 1988, at 2, col. 1.
8. *Id.*
9. J. Glover, *What Sort of People Should There Be?* 27 (1984).
10. *Id.*
11. *Id. See* Smith, Genetics, Eugenics and the Family: Exploring The Yin and The Yang, 8 U. *Tasmania L. Rev.* 4 (1984).
12. *Id.*
13. B. Stableford, *Future Man* 7 (1984).
14. J. Rifkin, *Algeny: A New Word—A New World* 160 (1983).
15. *Supra* note 9, at 38.
16. *Supra* note 13, at 161.

17. *Id.*
18. *Id.* at 162.
19. *Id.* at 161.
20. *Supra* note 13, at Ch. 9.
21. *Id.* at 158.
22. *Supra* note 4, at 45.
23. *Id.* at 46.
24. Thompson, Mapping the Human Genes, *Wash. Post Health Mag.,* Feb. 16, 1988, at Z8, col. 1.
25. *Id.*
26. *Time,* May 4, 1987, at 110.

    See also Gladwell, Harvard Scientists Will Patent for Genetically Altered Mouse: Award is First to be Issued for an Animal, *Wash. Post,* April 12, 1988, at 1, col. 1.
27. *Id.*
28. *Id.*
29. *Id.*
30. *Id.*
31. *Id.*
32. Schneider, A Patent on Life Forms Gets Genes Into Business, *Int'l. Herald Tribune,* June 9, 1987, at 1, col. 7.
33. *Id.*
34. *Id.*
35. *Id.*
36. Diamond v. Chakrabarty, 447 U.S. 303 (1980).

    The fear still persists that altered organisms might escape and cause considerable injury to local communities. A rural township in southern New Jersey passed an ordinance placing strict regulations on any outdoor testing of genetically engineered organisms within its boundaries, thereby making it the first municipality in the country to take a stand against the fledging biotechnology industry. *See* Gladwell, Towns Restricting Tests of Altered Organisms, *Wash. Post,* Mar. 20, 1988, at H5, col. 1.
37. *See* Boffey, Animal Rights, Fears of "Human Husbandry" Complicate Debate on Biotechnology, *Int'l. Herald Tribune,* June 10, 1987, at 6, col. 1; Current Topics, The Patenting of Animal Forms With New Traits, 61 *Australian L. J.* 324, 326 (July, 1987).

    The Office of Technology Assessment, together with the National Academy of Science, reported the risks associated with small-scale field tests of bio-engineered organisms are minimal.

See Gladwell, Report boosts biotechnology experiments: Risks of Small Scale Tests Outdoor Called Minimal, *Wash. Post*, May 5, 1988, at El, col. 2.

38. Stone, Knowledge, Survival and the Duties of Science, 23 *Am. Univ. L. Rev.* 231 (1973).
39. *Id.* at 232.
40. *Id.* at 235.
41. *Id.* at 234.
42. *Id.* at 236, 237.
43. *Id.*
44. *Id.*
45. *Id.*
46. *Id.* at 241.
47. *Id.* at 246.
48. *Id.*
49. *Id.* at 259.
50. *Id.*
51. *Id.* at 258.
52. *Id.* at 258.

A New South Wales Law Reform Commission Discussion Paper, *"In Vitro* Fertilization," found "no persuasive reason to restrict or prohibit total research on IVF embryos—provided the research was completed within a two-week period after the embryo was formed." The Law Reform Commission suggested the creation of a state advisory committee to assist physicians and scientists grappling with the vexatious ethical questions inherent within the area and stressed the need for self-regulation within the medical community. Mr. Russell Scott, the Deputy Chairman of the Law Reform Commission, summed up the medico-legal-scientific realities of the issues here by concluding, "Very strict legislation is not going to have any effect on IVF or which way research goes because it is happening on a global basis." O'Neill, Embryo IVF Research is Favoured, *Sydney Morning Herald*, July 30, 1987, at 1, col. 1. *See* New South Wales Law Reform Comm. *Artificial Conception, Discussion Paper 2, In Vitro Fertilization* (July 1987).

53. Kirby, Human rights—The Challenge of the New Technology, 60 *Australian L. J.* 170 (Mar. 1986).
54. *Id.* at 171.
55. *Id.* at 179. *See generally* R. Lillich & F. Newman, *International Human Rights: Problems of Law and Policy* (1979).

56. G.A. Res. 217A (III), U.N. *Doc* A/810 at 71 (1948).

57. *Supra* note 53, at 179.

58. *Supra* note 53, at 179.

59. Mount Isa Mines Ltd. v. Pusey, 125 C.L.R. 283, 395 (1970).

60. *See generally* Z. Cowen, *Reflections on Medicine, Bio-Technology and the Law* (1986). *See also* Barnett, Bio-technology—Can the Law Cope?, 15 *Anglo-American L. Rev.* 149 (1986); Smith, Bio-technology and the Law: Social Responsibility or Freedom of Scientific Inquiry, 31 *Mercer L. Rev.* 1 (1988).

CHAPTER 2

1. Compton, Science, Anti Science and Human Values, 1 *Amicus* 33 (1980).

2. Inventors Dream of Genes, *Time*, Oct. 20, 1980, at 72.

   The potential profits derived from manipulating the genetic code—be it either to create new forms of life sufficient to clean up toxic chemical wastes or to produce anti-cancer agents on a grant scale—spurred President Derek Bok of Harvard University to suggest that his University start its own genetic engineering firm. Strong faculty opposition, however, forced him to give up these plans. A Firm No, *Time*, Dec. 1, 1980, at 59.

   Office of Technology Assessment, U.S. Cong., *Commercial Biotechnology: An International Analysis* 542–546, (OTA-BA-218, Jan. 1984).

   See Cinelli, Biotechnological Research and Development: The Joint Venture as a Viable Corporate Entity in a High Risk Industry, 13 *J. Corp. L.* 549 (1988).

3. Diamond v. Chakrabarty, 447 U.S. 303 (1980).

4. Annas, Life Forms: The Law and The Profits, *Hastings Center Rep.* 21, 22 (Oct. 1978).

5. *Supra* note 1, at 37.

6. *Id.*

7. *See* Hilts, 'Rules' Drawn for Marketing Gene Research, *Wash. Post*, Mar. 28, 1982 at A1, col. 3; Will, The Spiral of Patents Pending, *Wash. Post*, June 22, 1980, at B7, col. 6.

8. Stone, Knowledge, Survival and The Duties of Science, 23 *Am. U. L. Rev.* 231 (1973).

9. *See generally* G. Smith, *Genetics, Ethics and the Law* 1, (1981).

10. J. Fletcher, *The Ethics of Genetic Control* 5 (1974).
11. Rivers, Genetic Engineering Portends a Grave New World, *Sat. Rev.* April 8, 1972, at 23. *See* Smith, Intimations of Immortality: Clones, Cryons and The Law, 6 U. *New So. Wales L. Rev.* 119 (1983); Smith, Beyond the Land of Oz: Clones, Cyborgs and Chimeras, 2 *Reps.* 6th *World Cong. Med. L.* 15 (1982).
12. *See generally* A. Toynbee, *Surviving the Future* (1971); *The Prospects of Western Civilization* (1949).
13. DNA is the basic genetic material that transmits inherited characteristics.
14. Clark, Begley & Hager, The Miracle of Spliced Genes, *Newsweek*, Mar. 17, 1980, at 62.

    See generally Baker & Clough, The Technological Use and Methodology of Recombinant DNA, 51 S. *Cal. L. Rev.* 1009 (1978); Berger, Government Regulation of the Pursuit of Knowledge: The Recombinant DNA Controversy, 3 *Sup. CT. L. Rev.* 83 (1978).
15. Scientists Want Limit Dropped on Gene Splitting Experiments, *Wash. Post*, Nov. 26, 1980, at C3, col. 5.

    But see Fields, Bizarre Circumstances Surround Chance Cloning of Banner Virus, *Chronicle of Higher Education*, Aug. 25, 1980, at 1, col. 1 (in violation of federal guidelines that bar genetic copying, a researcher at the University of California at San Diego cloned a virus); Holtzman, Patenting Certain Forms of Life: A Moral Justification, *Hastings Center Rep.* 9 (June 1979).

    See generally Deatherage, Scientific Uncertainty in Regulating Deliberate Release of Genetically Engineered Organisms: Substantive Judicial Review and Institutional Alternatives 11 *Harv. Envt'l L. Rev.* 203 (1987).
16. Neville, Philosophic Perspective on Freedom of Inquiry, 51 S. *Cal. L. Rev.* 1115, 1121 (1978).
17. Cohen, Restrictions of Research with Recombinant DNA: The Dangers of Inquiry and the Burden of Proof, 51 S. *Cal. L. Rev.* 1081, 1082, 1099 (1978).

    See Comment, Designer Genes That Don't Fit: A Tort Regime for Commercial Releases of Genetic Engineering Products, 100 *Harv. L. Rev.* 1086 (1987).
18. Fletcher, Ethics and Recombinant DNA Research, 51 S. *Cal. L. Rev.* 1131, 1139 (1978). Fletcher observes that there is nothing fundamentally unnatural or intrinsically wrong, or hazardous for the species, in the ambition that drives man to develop the

technology to understand himself. It would in fact seem more offensive to fail to use and develop man's natural curiosity and talen for asking questions or worse to try to suppress it.

19. *See* Toulmin, Science and Ethics: Can They be Reconnected: 73 U. *Chi. Mag.* 2 (1981).

20. *See* Roe v. Wade, 410 U.S. 113 (1973).

21. *See* J. Roslansky, *Genetics and the Future of Man* 46 (1966).

22. *See* G. Smith, *supra* note 9, at 2.

23. *See generally* R. Howard & J. Rifkin, *Who Should Play God?* (1977); Hilts, Genetic Scientist is Punished for Test Violations, *Wash. Post*, Mar. 23, 1981, at A1, col. 1.

24. Sinsheimer, Recombinant DNA—On Our Own, 26 *Bio-Science* 599 (1976).

25. Sinsheimer, Potential Risks, in *Research with Recombinant DNA* (Nat'l Academy of Science ed. 1977).

26. V. Goodfield, *Playing God* 71 (1977).

27. Fletcher, *supra* note 18, at 1138–1139.

28. *Id.* at 1138.

29. *See generally* T. Beauchamp & L. Walters, *Contemporary Issues in Bioethics* (1978); Smith, Uncertainties on the Spiral Staircase: Metaethics and The New Biology, 41 *Pharos Med. J.* 10 (1978).

30. *See* Irons & Sears, Patent 'Re-examination': A Case for Administrative Arrogation, 1980 *Utah L. Rev.* 287–288.

    By the Patent Clause, Congress is authorized "[t]o promote the Progress of Science and useful Arts, by securing for limited Times . . . Inventors the exclusive Right to their . . . Discoveries." U.S. Const. Art. I, § 8, col. 8.

31. *See* Sakraida v. AgPro, Inc., 425 U.S. 273, 279 (1976); Graham v. John Deere Co., 383 U.S. 1, 5–6 (1966); Atlantic Works v. Brady, 107 U.S. 192, 200 (1882)

    Interestingly, about 65–70% of litigated patents are invalidated. T. Beauchamp & I. Walters, *supra* note 29, at 305.

32. Diamond v. Chakrabarty, 447 U.S. 303 (1980).

33. Justice Brennan, writing in dissent, surveyed the Patent Act of 1793, as re-enacted in 1952, the Plant Patent Act of 1930, and the Plant Variety Protection Act of 1970 and concluded that there existed a strong congressional limitation against patenting bacteria. Id. at 322.

34. 35 U.S.C. § 101 (1976).

35. Diamond v. Chakrabarty, 447 U.S., at 307.

36. Gore, The Awesome Worlds Within a Cell, 150 *Nat'l Geographic* 355, 374–375 (1976).

37. Irons & Sears, Patents in Relation to Microbiology, 29 *Ann. Rev. Microbiology* 319, 331 (1975).

   See generally Kiley, Common Sense and the Uncommon Bacterium—Is 'Life' Patentable, 60 *J. Pat. Off. Soc'y* 468 (1978); Wegner, The Patentability of 'New' Manufacturers—The Living Invention, in *Patent Law ConferenceCoursebook* (Bureau of National Affairs ed. 1978).

   In a suit maintained by an individual who was successfully treated for leukemia by the University of California's Medical Center for the "commercial exploitation" of products medical researchers had obtained from his body tissue, the Court of Appeal of California held that human tissues and cells remain a person's property when taken for medical purposes and that the individual patient is entitled to share in the profits if they are used commercially. Moore v. The Regents of California, 249 *Cal. Rptr.* 494 (1988).

38. Application of Chakrabarty, 517 F. 2d 40 (C.C.P.A.) *dismissed* 439 U.S. 801 (1978), *rev'd sub nom.* Application of Bergy, 596 F.2d 952 (C.C.P.A.), cert. granted, 444 U.S. 924 (1979).

39. Student Papers, Microbiological Plant Patents, 10 *Idea* 87 (1966).

40. *Id. See* Cooper, Patent Protection for New Forms of Life, 38 *Fed. Bar. J.* 34 (1979); Kip, The Patentability of Natural Phenomena, 20 *Geo. Wash. L. Rev.* 371 (1952).

41. DeMott & Thomas, Test-Tube Life: Reg. U.S. Pat. Off., *Time*, June 30, 1980, at 52.

42. *See* Nelkin, Threats and Promises: Negotiating the Control of Research, 107 *Daedalus* 191 (1978).

43. 447 U.S. at 308 (1980).

44. *Id.*

45. *Id.* at 310.

   See generally Delgado & Miller, God, Galileo and Government: Toward Constitutional Protection for Scientific Inquiry, in 1 *Ethical, Legal and Social Challenges to a Brave New World* 231 (G. Smith ed. 1982).

46. 447 U.S. at 315.

47. *Id.* at 316–317.

   See Gladwell, Report Boosts Biotechnology Experiments: Risks of Small-Scale Tests Outdoors Called Minimal, *Wash. Post*, May 5, 1988, at El, col. 2.

48. *Id.*.

49. *Id.* at 317.

50. *Id.* at 311.
51. Sinsheimer, The Dawn of Genetic Engineering, 190 *Science* 768 (1975). *See* Fletcher, Moral Problems and Ethical Issues in Prospective Human Gene Therapy, 69 *Va. L. Rev.* 515 (1983).
52. *See* Smith, Manipulating the Genetic Code: Jurisprudential Conundrums, 64 *Geo. L. J.* 697 (1976).

    See also Office of Technology Assessment, *Impact of Applied Genetics* (1981); Note, Building a Better Bacterium: Genectic Engineering and the Patent Law After Diamond v. Chakrabarty, 81 *Colum. L. Rev.* 159 (1981).
53. Lederberg, Orthobiosis: The Perfection of Man in *Place of Value in a World of Facts* 29 (A. Tiselius & S. Nilsson eds. 1980).

CHAPTER 3

1. Ellis, Letting Defective Babies Die: Who Decides? 7 *Am. J. Law & Med.* 393, n. 1 (1981).
2. S. Hayes & R. Hayes, *Mental Retardation: Law, Policy and Administration* 48–49 (1981).
3. Childress, Triage in Neonatal Intensive Care: The Limitations of a Metaphor, 69 *Va. L. Rev.* 547 (1983).
4. *Id.*
5. *Time*, Aug. 9, 1982, at 42.
6. *Id.*
7. Pub. Law No. 93-112, tit. V, § 504, 87 Stat. 394 (1973).
8. 42 U.S.C. § 2000d (1976).
9. 20 U.S.C. § 1681 (1976).
10. *Supra* note 7.
11. *Id.*
12. 29 U.S.C. § 706(6) (1976).

    Since originally Section 504 did not grant either civil or criminal sanctions for violations lodged under it, in 1978 the Act was amended to make all those remedies which are available under The Civil Rights Act of 1964 applicable to Section 504 claims. Pub. Law No. 95-602, 92 Stat. 2955, Rehabilitation Comprehensive Services and Developmental Disabilities Amendments of 1978.
13. Gurmankin v. Costanzo, 411 F. Supp. 982 (E.D. Pa. 1976).
14. Camenisch v. Univ. of Texas, 616 F.2d 127 (5th Cir. 1980).

15. Ferris v. Univ. of Texas 558 F. Supp. 536 (W.D. Tex. 1983) (plaintiff sued to require the University to make its shuttle buses accessible to wheelchairs).
16. Discrimination Against the Handicapped by Withholding Treatment of Nourishment, 47 Fed. Reg. 26,027 (1982).
17. Nondiscrimination on the Basis of Handicap, 48 Fed. Reg. 9630 (1983).
18. *Id.*
19. U.S. v. Univ. Hospital, State Univ. of N.Y. at Stony Brook, Memo. Op. at pp. 43, 47.
20. 561 F. Supp. 395 (D.D.C. 1983).
21. *Id.*
22. 49 Fed. Reg. 1622 (Jan. 12, 1984).
23. American Hospital Assn. v. Heckler, 585 F. Supp. 541 (S.D.N.Y. 1984)
    See also The Economist, July 14, 1984, at 42; *Wash. Post*, Aug. 18, 1984, at A2, col. 3.
24. H. R. 1904.
25. S. 1003.
26. Report No. 98-159, House of Representatives, 98th Cong., 1st Sess., (1983).
27. 42 U.S.C. § 5101 *et seq.* (1984).
28. 49 Fed. Reg. 48170 (Dec. 10, 1984).
29. *Id.* at 48171.
30. *Id.* at 48170.
31. *Id.*
32. *Ark. Stat. Ann.* §§ 82-3801-3804 (Supp. 1978); *Ind. Code* § 35-1.58.5.7 (1975); New Mex. Stat. Ann. § 24.7.1 to 11 (1978).
    Following the principal Baby Doe Case in Indiana, three states enacted specific legislative programs designed to prevent an occurrence of this nature happening again: *Ariz. Rev. Stat. Ann.* § 36-2281.A (West Supp. 1983); *Ind. Code Ann.* § 31-6--4.3 (West Supp. 1983); and *La. Rev. Stat. Ann.* § 40:1299.36.1A (West Supp. 1983).
33. *Ark. Stat. Ann., supra.*
34. *New Mex. Stat. Ann., supra* note 32.
35. *Id.* at 24.7.3.
36. *Id.*
37. *Ind. Code, supra* note 32.
38. *Id.*
39. B. Sales, D. Powell & R. Van Duizend, *Disabled Persons and the*

Law 89–91 (1982).

See also Kuzma, The Legislative Response to Infant Doe, 59 *Ind. L. J.* 376 (1984).

40. 49 Fed. Reg. 1622 at 1652 (Jan. 12, 1984).

See generally Brief *Amicus Curiae* of Professor George P. Smith in Bowen v. American Hospital Association, U.S. Sup. Ct., Sept. 30, 1985, 476 U.S. 610 (1986).

41. *Id.*
42. *Id.* at 1653–1654.
43. *Id.* at 1652.
44. *Id.*
45. *Id.* at 1652.
46. *Id.*
47. *Id.* at 1651.
48. *Id.* at 1653–1654.
49. *Id.*
50. *Id.* at 1646–1649.
51. Marton, Fight for Baby Jane, *The Times*, Dec. 4, 1983, at 45, col. 1.
52. In re B (A Minor), C.A. 1981, 1 W.L.R. 1421.
53. *Id.* at 1422.
54. *Id.* at 1423.
55. *Id.* at 1424.
56. *Id.*

See Dickens, The Modern Function and Limits of Parental Rights, 97 *L. Q. Rev.* 462 (1981).

57. Kennedy, Reflections on The Arthur Trial, 59 *New Society* 13 (Jan. 7, 1982).
58. Glover, Letting People Die, *London Rev. of Books*, Mar. 4–17, 1982, at 3.
59. *Newsweek*, Int'l Ed., Dec. 12, 1982, at 27.
60. In the matter of William E. Webster, Guardian Ad Litem for Baby Jane Doe v. Stony Brook Hospital, et al, Oct. 28, 1983. Slip Opinion No. 676 at pp. 4, 5; 456 N.E. 2d 1186 C.N.Y. (1983).
61. Application of Frank T. Curio, M.D., 421 N.Y.S. 2d 965, 968 (1979).
62. *See* Van der Dussen, Givan: Court's "Doe" Action Set No Legal Precedent, *Bloomington Herald-Telephone*, April 23, 1982, at 1, col. 1.

In re Infant Doe, No. GU 8204-00 (Cir. Ct. Monroe County, Ind., April 12, 1982) writ of mandamus dismissed *sub nom.* Infant Doe v. Baker, No. 482—S140 (Ind. Sup. Ct., May 27, 1982) (case mooted by child's death).

63. Robertson, Dilemma in Danville, *Hastings Center Rep.*, 5, (Oct. 1981).
64. See The Moral Dilemma of Siamese Twins, *Newsweek*, June 22, 1981, at 40.
65. State Drops Siamese Twins Case, *N.Y. Times*, Apr. 17, 1983, at 6, col. 2.
66. Campell & Duff, Moral and Ethical Dilemmas in the Special Case Nursery, 289 *New Eng. J. Med.*, 890 (1973).
67. Ellis, Letting Defective Babies Die: Who Decides? 7 *Am. J. Law & Med.* 393, 399 (1981); Robertson, Involuntary Euthanasia of Defective Newborns: A Legal Analysis, 21 *Stan. L. Rev.* 213 (1974).
68. M. Shapiro & R. Spece, Jr., *Cases, Materials and Problems on Bioethics and Law*, 702, 703 (1981).
69. *Supra* note 66. Harvard, Legislation is Likely to Create More Difficulties Than It Resolves, 9 *J. Med. Ethics* 18 (1983).
70. *Supra* note 66, at 893–894.
71. Fost, Counseling Families Who Have a Child with a Severe Congenital Anomaly, 1981 *Pediatrics* 321–323; Editorial, Severely Handicapped Infants, 7 *J. Med. Ethics* 115 (1981); Diamond, *Treatment Versus Non-Treatment for the Handicapped in Infanticide and the Handicapped Newborn* at 55, 62 (D. Horan & N. Delaboyd eds. 1982).
72. Koop, Ethical and Surgical Consideration in the Care of Newborns with Congenital Abnormalities in D. Horan & N. Delaboyd *supra*, at 71.
73. *Id.*
74. The President's Commission for The Study of Ethical Problems in Medicine and Biomedical and Behavioral Research, *Deciding to Forego Life-Sustaining Treatment: A Report on Ethical, Medical and Legal Issues on Treatment Decisions* (Committee Print, Mar. 21, 1983) at 226, 228. This Commission ceased operating in 1983.
75. *Id.*
76. *Id.*
77. *Current Opinions of the Judicial Council of the American Medical Association*, A.M.A., 1982, at 9.
78. *Euthanasia, Aiding Suicide and Cessation Treatment Report of the Law Reform Commission of Canada*, No. 20 at 24–26 (1983).
79. Sternberg, Lying Hopelessly Ill, Infant Tests Law of Hospital Survival, *Wash. Post*, Apr. 3, 1983, at A4, col. 1.

80. Stedman's *Medical Dictionary*, at 1476 (W. Cornett ed. 1976).

81. Childress, Triage in Neonatal Intensive Care: The Limitations of a Metaphor, 69 *Va. L. Rev.* 547, 549 (1983).

82. See D. Rund & T. Rausch, *Triage* 3–10 (1981).

83. G. Winslow, *Triage and Justice* 1 (1982).

84. *Id.* at 2.

85. *Id.*

86. *Id.* at 5.

87. *Supra* note 81, at 550.

88. *Id.*

89. Note, Scarce Medical Resources, 69 *Colum. L. Rev.* 620 (1969). See generally, Smith, *Triage*: Endgame Realities, 1 *J. Contemp. Health L. & Pol'y.*, 143 (1985).

90. *Supra* note 83, at 106.

91. *Id.* at 63–86.

92. *Id.* at 87.

93. *Id.*

94. *Id.*

95. *Id.* at 88–106.

96. Gewrith, Can Utilitarianism Justify Any Moral Rights? in *Ethics, Economics and the Law*, 167 (J. Pennock & J. Chapman eds. 1982).

97. *Id.* at 167, 168.

98. *Id.* at 168.

99. *See generally* Smith, Uncertainties on the Spiral Staircase: Metaethics and The New Biology, 41 *The Pharos J.* 10 (1977).

100. G. Smith, *Genetics, Ethics and the Law*, 2, 8, (1981).

101. *Id.*

102. *Id.*

103. Young, Infanticide and The Severely Defective Infant, in *Problems of the Handicapped at* 131 (R. Laura ed. 1980). *See also* M. Tooley, *Abortion and Infanticide* (1983).

104. *Id.*

105. Editorial, Bioethics and The Law, *Reform* (The Law Reform Commission J.), at 10 (Jan. 1979).

106. Cranford & Doudera, The Emergence of Institutional Ethics Committees, L., *Med. & Health Care* 13 (Feb. 1984).

107. In re Quinlan, 355 A. 2d 647 (N.J. 1976).

108. *Id.*

109. *Supra* note 62.

110. Weber v. Stony Brook, 456 N.E. 2d 1186 (N.Y. 1983).

111. 49 Fed. Reg. 1622 (Jan. 12, 1984).

112. *Supra* note 106, at 14.
113. *Id.* at 14, 15.
114. *Id.* at 17.
115. *Id.*
116. *Id.*
117. *Id.* at 15.
118. *Id.* at 16.
119. *Id.*
120. *Id.* at 19.
121. *Id.*
122. G. Smith, *Genetics, Ethics and the Law*, 2, 8 (1981).
123. Kennedy, Reflections on the Arthur Trial, 59 *New Society* 13 (Jan. 7, 1982).
124. Fletcher, Indicators of Humanhood: A Tentative Profile of Man, *Hastings Center Rep.* 1 (Nov. 1972).
125. R. McCormick, To Save or Let Die: The Dilemma of Modern Medicine in *How Brave a New World*, at 339 (R. McCormick ed. 1981).
126. R. Veatch, *A Theory of Medical Ethics* 244 (1981).
127. *Id.* at 245.
128. Lorber, Early Results of Selective Treatment of Spina Bifida Cystica, 4 *Br. J. Med.* 201, 203 (1973).
129. Draft, Neonatal Intensive Care Unit Admission Guidelines, Department of Health, New South Wales, Australia, 1984.

    At the First Conference of Experts of the German Society for Medical Law held at Einbeck, West Germany, June 27–29, 1986, recommendations were developed that set limits on the duty to provide medical care for seriously defective newborns. In cases where severe, nonremedial deficiencies exist and death is certain (e.g., inoperable heart defects), and can only be postponed, or where the newborn will never have the possibility of communicating with his environment (e.g., microcephaly) or where the vital functions are only maintainable for any length by means of intensive medical intervention, the physician should *not* afford all available means for medical treatment to his patient.
130. Chervenak, Farley, Walters, Hobbins, & Mahoney, When IS Termination of Pregnancy During the Third Trimester Morally Justifiable?, 318 *New Eng. J. Med.* 591 (1984).
131. *Id.*
132. *Id.* at 502.

133. *Id.*
134. *Id.* at 503.
135. Young, Caring for Disabled Infants, *Hastings Center Rep.* 15 (Aug. 1983).
136. *Id.* at 16.
137. *Id.*
138. Strong, The Tiniest Newborns, *Hastings Center Rep.* 14 (Feb. 1983).
139. E. Kuge, *The Practice of Death* 191 (1975).
140. *Supra* note 135.
141. P. Ramsey, *The Patient as Person* viii (1970).
142. *Supra* note 100.
143. Dickens, The Modern Function and Limits of Parental Rights, 97 *L. Q. Rev.* 462 (1981).
144. Burger, Reflections on Law and Experimental Medicine in 1 *Ethical, Legal and Social Challenges to a Brave New World* at 211 (G. Smith ed. 1982).
145. 50 Fed. Reg. 14878 (April 15, 1985).
146. *Id.* at 14881, 14883.
147. *Id.*
148. *Id.*
149. *Id.*
150. Murray, The Final, Anticlimatic Rule on Baby Doe, *Hastings Center Rep.* 5, 8 (June 1985).
151. *Id.* at 8.
152. *Supra* note 145, at 14889.
153. *Id.*
154. *Supra* note 150, at 8.
155. *Id.*
156. *Id.*
157. *Id.*
158. Bowen, Sec. of Health & Human Services v. American Hosp. Assoc., 476 U.S. 610, (1986).
159. Ala. Code §§ 26-15-1–26-15-4 (1975); Alaska Stat. § 47.10.010 (1962); Ariz. Rev. Stat. Ann. § 13.3673 (1956); Ark. Stat. Ann. § 42-807 (1977); Cal. Penal Code § 11165 (West 1980); Colo. Rev. Stat. § 18-6-401 (1973); Conn. Gen. Stat. § 53-20 (1949); Del. Code Ann. tit. 31, § 301 (1974); D.C. Code Ann. § 6-2101 (1981); Fla. Stat. § 827.07 (1981); Ga. Code Ann. 74-111 (1933); Hawaii Rev. Stat. § 3501-1 (1976); Idaho Code § 16-1601 (1949); Ill. Rev. Stat. Ch. 23, § 2368 (1977); Ind. Code Ann. § 31-6-11-1 (Burns

1980); Iowa Code § 235 (1979); Kan. Stat. Ann. § 38-704 (1965); Ky. Rev. Stat. § 15-900 (1984); La. Rev. Stat. Ann. § 14.93 (West 1970); Me. Rev. Stat. Ann. Tit. 19, § 218 (1964); Md. Pub. Safety Code Ann. § 35A (1973); Mass. Gen. Laws Ann., Ch. 273, § 1 (West 1975); Mich. Comp. Laws Ann. § 722.622 (West 1983); Minn. Stat. Ann. § 626.556 (West 1962); Miss. Code Ann. § 43.19.1 (1972); Mo. Rev. Stat. § 210.110–210.165 (1982); Mont. Code Ann. § 10-503 (1977); Neb. Rev. Stat. § 28-707 (1979); Nev. Rev. Stat. §§ 200.501–200.509 (1973); N. H. Rev. Stat. Ann. § 169.38 (1955); N.J. Stat. Ann. § 26.24-4 (West 1982); N. M. Stat. Ann. § 30-6-1(c) (1978); N. Y. Social Services Law § 412 (McKinney 1983); N. C. Gen. Stat. § 14-318.2 (1981); N. D. Cent. Code § 50-25.1-02 (1982); Ohio Rev. Code Ann. § 29-1922 (1982); Okla. Stat. Ann. Tit. 21, § 843 (West 1983); Ore. Rev. Stat. § 418.770 (1983); Pa. Stat. Ann. Tit. 11, § 23 (Purdon 1965); R. I. Gen Laws § 40-11-2 (1956); S. C. Code Ann. § 20-7-10 (Law. Co-op 1976); S. D. Codified Laws Ann. § 26-10-10 (1984); Tenn. Code Ann. § 37-1200 (1977); Tex. Family Code Ann. § 35.04 (Vernon 1975); Utah Code Ann. § 76-5-109 (1953); Vt. Stat. Ann. Tit. 13 § 2801 (1974); Va. Code § 63.1.248 (1950); Wash. Rev. Code § 26.44.020 (1974); W. Va. Code § 49-1-3 (1980); Wis. Stat. Ann. § 940.201 (West 1982); Wyo. Stat. § 6-4-504 (1977).

160. Pub. L. No. 98-457, 98 Stat. 1749, 42 U.S.C. §§ 201 et. seq. (1984). See also, C.F.R. § 1340.15 (1985).

161. 45 C.F.R. § 1340.15(b)(2) (1985).

162. *Id.*

Final regulations implementing the new requirements were published by the Department of Health and Human Services April 15, 1985. *See* 50 Fed. Reg. 14878 (1985).

163. Kopp & Grant, The "Small Beginnings" of Euthanasia: Examining the Erosion in Legal Prohibitions Against Mercy-Killing, 2 *Notre Dame J. L. Ethics & Pub. Pol'y* 585, 618 (1986).

CHAPTER 4

1. *See* S. Stanley, *The New Evolutionary Timetable* (1981): T. Dobzhansky, *Genetic Diversity and Human Equality* (1973); Muller, The Human Future in *The Humanist Frame* 401 (J. Huxley ed. 1961); Muller, Human Values in Relation to Evolution, 127 *Science* 625–629 (Mar. 21, 1958).

2. Gorney, The New Biology and The Future of Man, 15 *U.C.L.A. L. Rev.* 273, 291 (1968).

3. Robinson, Genetics and Society, 1971 *Utah L. Rev.* 487.

Approximately 30,000 severely defective infants are born each year and afflicted with grave handicapping conditions that range from spina bifida to anencephaly. Ellis, Letting Defective Babies Die: Who Decides? 7 *Am. J. L. & Med.* 393, n. 1 (1981).

4. *Supra* note 1.

5. Waltz & Thigpen, Genetic Screening and Counseling: The Legal and Ethical Issues, 68 *Nw. U. L. Rev.* 696–698 (1973).

6. *Id.* at 698.

7. Kass, The New Biology: What Price Relieving Man's Estate? 174 *Science* 779, 780 (1971).

See also C. Heintze, *Genetic Engineering: Man and Nature in Transition* (1973).

See R. Blank, *The Political Implications of Human Genetic Technology* 66 *passim* (1983).

8. *See generally* Symposium—Reflections on the New Biology, 15 *U.C.L.A. L. Rev.* 267 (1968).

Creative, scientific impulses for research and investigation should be neither systemized nor controlled. "Some part of life—perhaps the most important part—must be left to the spontaneous action of individual impulse, for where all is system, there will be mental and spiritual death." B. Russell, *The Impact of Science on Society* 89 (1952).

See R. Blank, *The Political Implications of Human Genetic Technology* 66 *passim* (1983).

9. Waltz & Thigpen, *supra* note 5, at 696.

See also M. Frankel, *Genetic Technology: Promises and Problems* (1973); Fletcher, Ethics and Recombinant DNA Research, 51 *S. Cal. L. Rev.* 1311 (1978).

10. *See* Vukowich, Ch. 3, The Dawning of the Brave New World— Legal, Ethical and Social Issues of Eugenics, in 2 *Ethical, Legal and Social Challenges to a Brave New World* (G. Smith ed. 1982).

11. Frankel, The Spector of Eugenics, 57 *Commentary* 25, 30 (1974).

12. *Id.*

To be justifiable, the acceptance or rejection of eugenic policies should be based upon more than one criterion. The following requisites should be a part of every eugenic program: scientific validity (e.g., a demonstration of sufficient genetic variation to allow for selection of the attribute in question);

moral acceptability (i.e., a demonstration that the attributes chosen for selection are properly considered socially desirable); and ethical acceptability (i.e., a demonstration that the programs needed to institute a eugenic program do not compromise individual rights and liberties presently sanctioned by both public policy and the law). Lappe, Why Shouldn't We Have Eugenic Policy?, in *Genetics and the Law* 421 at 425 (A. Milunsky & G. Annas eds. 1976).

See also Osborn, Qualitative Aspects of Population Control: Eugenics and Euthenics, 25 *Law & Contemp. Probs.* 406 (1960).
13. Smith, Through a Test Tube Darkly: Artificial Insemination and the Law, 67 *Mich. L. Rev.* 127, 147 (1968).
14. T. Dobshanzky, *Mankind Evolving* 245 (1962); M. Haller, *Eugenics* 3 (1963).

See also Green, Genetic Technology: Law and Policy for the Brave New World, 48 IND. L. J. 4559 (1973): Dobzhansky, Comments on Genetic Evolution, 90 DAEDALUS 451, 470–473 (1961); Tooley, Abortion and Infanticide, 2 J. *Philosophy & Pub. Affairs* 37 (1972).
15. See generally, Hafen, Constitutional Status of Marriage, Kinship, and Sexual Privacy—Balancing the Individual and Social Interests, 81 *Mich. L. Rev.* 463 (1983); Symposium, The Family in Legal Transition, 1983 *U. Ill. L. Rev.* 99.
16. Plato, *The Republic*, bk. 5 at 166–70 (J. Davies & D. Vaughn trans. 1891).
17. C. Darwin, *Descent of Man and Selection in Relation to Sex* 402–403 (1871).
18. Comment, Eugenic Artificial Insemination: A Cure for Mediocrity?, 94 *Harv. L. Rev.* 1850, 1852 (1981).
19. F. Galton, *Heredity Genius* 1 (1869).
20. *Supra* note 18, 1852.
21. Cynkar, Buck v. Bell: Felt Necessities v. Fundamental Values?, 81 *Colum L. Rev.* 1416, 1420 (1981).
22. *Id.*

See generally G. Stine, *Biosocial Genetics: Human Heredity and Social Issues* (1977).
23. *Id.* at 1421.

See generally V. McKusick, *Mendelian Inheritance in Man* (1978).
24. *Id.* at 1422–1425.

See also J. Fletcher, *Coping with Genetic Disorders* 3–32 (1982).

25. *Id.*
26. *Id.* at 1428.
27. *Id.*
    Beckwith, Social and Political Uses of Genetics in the United States: Past and Present, 265 *Annals of the New York Academy of Sciences* 46 (1976).
28. H. Laughlin, *The Legal Status of Eugenical Sterilizations* 65 (1929).
29. *Id.* at 60.
    See Lappe, Moral Obligations and the Fallacies of Genetic Control, 33 *Theological Studies* 411 (1972).
30. *Supra* note 21, at 1433.
31. *Id.* at 1434.
32. *Id.* at 1454.
33. *Id.* at 1455.
34. *Id.* at 1456.
35. Davis, Ethical and Technical Aspects of Genetic Intervention, 285 *New Eng. J. Med.* 799 (1977).
36. *Supra* note 5, at 700.
37. *Id.*
    See also Kobrin, Confidentiality of Genetic Information, 30 *U.C.L.A. L. Rev.* 1283 (1983).
38. Robinson, Genetics and Society, 1971 *Utah L. Rev.* 487 at 488 n. 24.
39. *Id.* at 48.
40. *Id.*
    See Ramsey, Screening: An Ethicist's View, in *Ethical Issues in Human Genetics* 147 at 154 (B. Hilton, D. Callahan, M. Harris, P. Condliffe & B. Berkley, eds. 1973); Lappe et al, Ethical and Social Issues in Screening for Genetic Disease, 286 *New Eng. J. Med.* 1129 (1972).
41. *See* Note, A Cause of Action for Wrongful Life, 55 *Minn. L. Rev.* 58 (1970); Annot., 22 A. L. R. 3d 1441 (1968).
42. Rivers, Grave New World, *Sat. Rev.*, April 8, 1972, at 23, 26.
    There are four areas in which genetic disease may be classified: single gene effects; chromosomal abnormalities; congenital malformation; and serious constitutional disorders. The incident of single gene effects—of which the most commonly known are phenlketonuria (P.K.U.), Tay-Sachs disease and X-linked mental retardation—is 11.2 affected births per 1000 births. Chromosomal abnormalities—which would include Down's syndrome and Turner's syndrome—account for 5.4 per

1000 births. The incidence of congenital malformation is 14.1 per 1000 births and the serious constitutional disorders—which include diabetes and epilepsy—occur in 14.8 per 1000 births. S. Hayes & R. Hayes, *Mental Retardation: Law, Policy and Administration* 28, 29 (1982). See also, G. Roderick, *Man and Heredity* 225 (1968); H. Papazian, *Modern Genetics* 77 (1967); S. Scheinfeld, *Your Heredity and Environment* 189 (1965).

43. Walters, Introduction to Genetic Intervention and Reproduction Technologies, in *Contemporary Issues in Bioethics* at 567 (T. Beauchamp & L. Walters eds. 1978).

   See Nelson, Swint & Caskey, An Economic Evaluation of a Genetic Screening Program for Tay-Sachs Disease, 30 *Am. J. Human Genetics* 160 (1978).

44. National Sickle Cell Anemia Control Act, 42 U.S.C. § 3306 *et seq.* (Supp. III, 1973). A Cerami, E. Washington, *Sickle-Cell Anemia* (1974).

   See also A. Etzioni, *Genetic Fix* 132 (1973).

   See Reilly, Government Support of Genetic Services 25 SO-CIAL BIOLOGY 23 (1978); Culliton, Cooley's Anemia: Special Treatment for Another Ethnic Disease, 178 *Science* 593 (1972); National Cooley's Anemia Control Act (Public Law 92-414 (1972)). There has also been special congressional concern over the study and regulation of Huntington's chorea (89 Stat. 349 (1975)) and hemophilia (90 Stat. 350 (1975)).

45. *See, e.g., Ill. Ann. Stat.* ch. 122 § 27–28 (Smith-Hurd Supp. 1979) (exception for refusal of physical examination on constitutional grounds); *Mass. Gen. Laws Ann.* ch. 76 § 15A (Supp. 1979) (mandatory only if child susceptible); N.Y. EDUC. § 904 (McKinney Supp. 1978–79) (exception for refusal based on religious beliefs). *See also Va. Code Ann.* §§ 32-112.20 to 112.23 (Supp. 1979) (voluntary screening program).

   Dr. Linus Pauling has suggested that sickle-cell anemia carriers be identified by tattooing the forehead of every carrier. Other recessive genes, such as hemophilia and phynylketonuria, could be similarly identified. See Pauling, Foreword, Symposium—Reflections on the New Biology, 15 *U.L.A. L. Rev.* 267, 270 (1968).

   Today, some 43 states have PKU screening laws; another 14 test neonatally for a variety of screening problems other than PKU. Reilly, State Supported Mass Genetic Screening Programs, in *Genetics and the Law* 159, 164 (A. Milunsky & G. Annas eds. 1976).

46. *N.Y. Dom. Rel. Law* § 13-aa (McKinney 1977). Other states provide for voluntary premarital testing for sickle cell anemia. *See* CAL. HEALTH & SAFETY CODE §§ 325-331 (West Supp. Pamp. 1978); *Ga. Code Ann.* § 53-216 (1974).

47. *See Va. Code Ann.* § 32-122.22 (Supp. 1979). *See* Antley, Variables in the Outcome of Genetic Counseling, 23 *Soc. Biology* 108 (1976).

    A genetic counselor "has freedom to persuade, according to his personal convictions, but he does not have freedom to coerce, based upon his inherent power in the counseling milieu. He must accept the counselee as the ultimate decision maker." Shaw, *Genetic Counseling in Human Genetics: Readings on the Implications of Genetic Engineering* 199 at 200 (T. Mertens ed. 1975).

48. Waltz & Thigpen, Genetic Screening and Counseling: The Legal and Ethical Issues, 68 Nw. U. L. Rev. 696 at 701–702, nn. 28, 29 (1973).

    See also Screening and Counseling for Genetic Conditions: A Report on the Ethical, Social and Legal Implications of Genetic Screening, Counseling, and Education Programs, President's Commission for the Study of Ethical Problems in Medicine and Behavioral Research (Feb. 1983); J. Fletcher, *Coping with Genetic Disorders* 50–74 (1982).

49. Waltz & Thigpen, *supra*, at 702–702, nn. 30–31.

    Confusion as to the significance of possessing the defective gene not only renders screening programs less effective in discouraging reproduction, but the failure to differentiate between the disease and the trait also increase the sitmatization to which carriers are subjected. *Id.*

50. *Supra* note 11, at 30.

51. *Id.*

52. *Id.*

    While the United States Air Force ended its ban on maintaining cadets at the Academy who were carriers of sickle-cell anemia in 1981, it has been reported that some six or more major American corporations endeavor to screen prospective employees for genetic deficiencies (and particularly their sensitivity to toxic substances). In 1982, nearly five dozen other Fortune 500 firms reported that within five years they too expected to follow a similar policy. Kevles, Annals of Eugenics, *The New Yorker* 116, 117 (Oct. 29, 1984).

53. *Supra* note 48, at 712.

    See Isaacs, The Law of Fertility Regulation in the United States: A 1980 Review, 19 *J. Fam. L.* 65 (1980–1981).
54. *Id.* at 711–712.
55. *Id.*
56. Cf. Schmerber v. California, 384 U.S. 757, 772 (1966) (compulsory blood test to determine intoxication of automobile driver not unreasonable search).
57. Vukowich, *supra* note 10, at 208.
58. Pauling, *supra* note 45, at 270–271.
59. *See* Vukowich, *supra* note 10, at 215–216.
60. *Id.* at 216.
61. *Id.*
62. *See,* e.g., Eisenstadt v. Baird, 405 U.S. 438, 452–455 (1972) (forbidding—on morality grounds—sale or gift of contraceptives to unmarried persons conflicts with fundamental constitutional rights); Loving v. Virginia, 388 U.S. 12 (1967) (state may not infringe freedom to marry person of another race); Griswold v. Connecticut, 381 U.S. 479, 481–486 (1965) (statute forbidding use of contraceptives violates constitutionally protected right of marital privacy).
63. Skinner v. Oklahoma, 316 U.S. 535, 541 (1941).

    Concurring in Griswold v. Connecticut, Justice Goldberg commented that a compulsory birth control law unjustifiably would abridge the constitutional rights of marital privacy. 281 U.S. 479, 497 (1965) (with Warren, C.J. & Brennan, J. concurring).
64. *See* Roe v. Wade, 419 U.S. 113, 153 (1973).
65. Golding & Golding, Ethical and Value Issues in Population Limitation and Distribution in the United States, 24 *Vand. L. Rev.* 495, 511 (1971).
66. *Id.* at 512. The authors conclude, however, that the unrestricted freedom to procreate should be abridged only for "good of momentous order." *Id.*
67. 274 U.S. 200 (1927).
68. *Id.* at 207.

    Justice Holmes, speaking for the Court, stated: "We have seen more than once that the public welfare may call upon the best citizens for their lives. It would be strange if it could not call on those who already sap the strength of the State for these lesser sacrifices, often not felt to be such by those concerned, in order to prevent our being swamped with incompetence. It is better for all the world, if instead of waiting to execute degenerate offspring for crime, or to let them starve for their imbecility, society can prevent those who are manifestly unfit from continuing their kind." *Id.*

See also In re Sterilization of Moore, 289 N.C. 95, 221 S.E. 2d 307 (1976).

69. The present eugenic sterilization statutes are: *Cal. Penal Code* § 645 (West 1970); *Del. Code Ann.* tit. 16, § 5701 (1983); *Idaho Code* §§ 39-3901–3910 (1985); *Me. Rev. Stat. Ann.* tit. 34B §§ 7001–7017 (Supp. 1985); *Minn. Stat. Ann.* § 252A.13 (1982); *Miss. Code Ann.* §§ 41-45-1-41-45-19 (1981 & Supp. 1985); *Mont. Code Ann.* §§ 50-5-501-50-5-505 (1985); *N.C. Gen. Stat.* §§ 35-36-35-50 (1984); *Or. Rev. Stat.* § 436.205–436.335 (1983); *S.C. Code Ann.* §§ 44-47-10-44-47-100 (1985); *Utah Code Ann.* §§ 64-10-1-64-10-16 (1986); *Vt. Stat. Ann.* tit. 18, §§ 8701–8716 (1968 & Supp. 1985); *Va. Code* §§ 27-16-1-27-16-5 (1976). It has been estimated that over 70,000 people have been sterilized under such statutes. *Statistics from Human Betterment Ass'n of America, Summary of U.S. Sterilization Laws* 2 (1958).

One should distinguish these eugenic sterilization statutes from those sterilization statutes which are wholly voluntary in nature. Among the voluntary statutes are: *Or. Rev. Stat.* § 435.305 (1983); *N.M. Stat. Ann.* §§ 24-1-14, 24-9-1 (1984); *Ga. Code Ann.* §§ 84–932 (1985); *N.C. Gen. Stat.* §§ 90-271-90-275 (1985). These statutes are essentially contraceptive and therapeutic and not eugenic in nature.

70. *See*, e.g., Oregon v. Cook, 9 Ore. App. 224, 230, 495 P.2d 768, 771–772 (1972) (equal protection challenge based on indigency rejected); In re Cavitt, 182 Neb. 712–721, 157 N.W. 2d 171, 178 (1968), appeal dismissed, 396 U.S. 996 (1970).

See also Dunn, "Eugenic Sterilization Statutes: A Constitutional Re-evaluation," 14 *J. Fam. L.* 280 (1975).

71. *See* Shapiro v. Thompson, 394 U.S. 618, 638 (1969).

72. *See* Oregon v. Cook, 9 Oreg. App. 224, 230, 495 P.2d 768, 771–727 (1972).

73. Buck v. Bell, 274 U.S. 200, 208 (1927).

The Court's rationale acquires additional significance because it became the basis for distinguishing Buck in the case of Skinner v. Oklahoma—where the High Court invalidated a statute providing for the sterilization of habitual criminals. The

Skinner Court concluded that the questioned statute violated the fourteenth amendment's equal protection clause. See 316 U.S. 535, 542 (1941).

74. The statute challenged in Buck required only that experience demonstrate heredity plays an important role in the transmission of the mental defect. *See* 274 U.S. at 206. The inmate involved, however, was the daughter of a feebleminded mother. *Id.* at 205.

See generally Murray, "Marriage Contracts for the Mentally Retarded," 21 *Cath. Law.* 182 (1975); Ferster, Eliminating the Unfit—Is Sterilization the Answer? 27 *Ohio St. L. J.* 591 (1966).

75. *See* Waltz & Thigpen, *supra* note 48, at 721, n. 131.

76. *Supra* note 57.

See also Baron, Voluntary Sterilization of the Mentally Retarded, in *Genetics and the Law* 267 (A. Milunsky & G. Annas eds. 1976).

77. 274 U.S. at 207.

Unrestricted genetic transmission forces a heavy burden upon society. The Juke and Kallikak family histories reveal clearly this point. Max Juke resided in Ulster County, New York. He had two sons who married two of six sisters of a local feebleminded family. One other sister left the area; the other three married mental defectives. From these five sisters, 2094 direct descendants and 726 consortium descendants were traced by 1915 into 14 states. All of them were feebleminded and the cost to society from their welfare payments, illicit enterprises, jail terms, and prostitution brothels was $2,516,685.

Martin Kallikak, Sr., fostered a son, Martin Jr., by a feebleminded girl during the Revolutionary War. Martin Jr. married a feebleminded girl and they in turn had seven children, five of whom were similarly afflicted. From these progeny sprung 480 descendants, 143 feebleminded, 46 normals, and 291 of unknown mental stature. J. Wallin, *Mental Deficiency* 43–44 (1956).

Various estimates have been made relative to the lifetime costs of various genetic diseases—often with rather astonishing results. For example, it has been calculated that the lifetime costs of maintaining a seriously defective individual is $250,000; this assumes, of course, institutionalization. Conservative estimates place the number of new cases of Down's syndrome in the United States at 5000 each year, or one in every 700 live births. Using the $250,000 figure for the cost of maintenance, the lifetime committed expenditure for new cases of Down's syndrome alone comes to at least $1.25 billion yearly, admittedly a staggering figure for but one disease entity.

Another way of calculating the toll of genetic disease is to estimate the future life years' cost. One widely cited estimate indicates that some 36 million future life years are lost in the United States by birth defects, putting the figure for recognized genetic disease (80% of birth defects being genetic in whole or in part) at 29 million future life years lost, or several times as much as from heart disease, cancer, and stroke. WHAT ARE THE FACTS ABOUT GENETIC DISEASE? at 27, 29, U.S. Dept. of H.E.W., Public Health Service, N.I.H., DHEW Pub. No. (NIH) 75-370 (1978). *See also* M. Frankel, *Genetic Technology: Promises and Problems* 46–77 (1973).

78. S. Hayes & R. Hayes, *Mental Retardation: Law, Policy and Administration* 73 (1982).
79. *Id.*
80. *Id.* at 76.
81. *Id.*
82. *Id.* at 31.
83. *Id.*
84. *Id.* at 32.
85. *Id.* at 30.
86. *Id.*
87. *Id.*

In Victoria, there are 43 notifiable diseases under the Health Act—but these do not include genetic abnormalities identifiable in newborns and there is, furthermore no compulsion for treatment. *Id.*

88. *Id.* at 33.
89. *Id.*
90. *Id.*
91. *Id.*
92. *Id.*
93. *Id.* at 48–49.
94. Smith, Through A Test-Tube Darkly: Artificial Insemination and the Law, 67 *Mich. L. Rev.* 127, at 148 (1968).

It is generally agreed that it is best for any AID baby not to know of its origins. The donor should not be told if his donation of semen resulted in a successful impregnation and birth. Attalah, Report from A Test Tube Baby, *N. Y. Times Mag.*, April 18, 1976, 16 and 17, 51.

95. Smith, *supra*, at 145, 146.

The Repository for Germinal Choice became operational in 1979 in Escondido, California, and is designed to make available the sperm of Nobel Prize winners and other "creative, intelligent people." *See Playboy* Interview: William Schockley, 27 *Playboy* 69 (Aug. 1980).

See also Broad, A Bank for Nobel Sperm, 207 *Science* 1326 (Mar. 1980).

96. *Id.*

97. *Id.*

See generally S. Pickens, *Eugenics and the Progressive* (1968); Medawar, The Genetic Improvement of Man, 4 *Australasian Annals of Med.* 317 (1969).

98. Brewer, Eutelegenesis, 27 *Eugenics Rev.* 121, 123, 126 (1935).

See generally Smith, The Razor's Edge of Human Bonding: Artificial Fathers and Surrogate Mothers, 4 W. *New Eng. L. Rev.* 639 (1983).

99. Vukowich, *supra* note 10, at 230–231.

100. Smith, Sexuality, Privacy and the New Biology, 67 *Marq. L. Rev.* 263 (1984).

101. Rovrik, The Embryo Sweepstakes, *N.Y. Times Mag.*, Sept. 15, 1974, at 17.

102. *Id.*

103. *Id.*

104. *Time*, July 24, 1978, at 47.

105. *See Test-Tube Babies: A Guide to Moral Questions, Present Techniques and Future Possibilities* (W. Walters & P. Singer eds. 1982); Edwards & Steptoe, Current Statutes of *In Vitro* Fertilization and Implantation of Human Embryos, *The Lancet* 1265 (Dec. 3, 1983); Biggers, *In Vitro* Fertilization and Embryo Transfer in Human Beings, 304 *New Eng. J. Med.* 336 (1981).

106. *Making Babies: The Test Tube and Christian Ethics* (A. Nichols & T. Hogan eds. 1984); Symposium, *In Vitro* Fertilization: The Major Issues, 9 *J. Med. Ethics* 192 (1983).

See also Annas & Elias, *In Vitro* Fertilization and Embryo Transfer: Medico Legal Aspects of a New Technique to Create a Family, 17 *Fam. L. Q.* 199 (1983).

107. Gaylin, We Have the Awful Knowledge to Make Exact Copies of Human Beings, *N.Y. Times Mag.*, Mar. 5, 1972, 11 at 48.

108. *See* Gaylin, *supra*, at 48; cf. Rorvik, supra note 101 at 50 (eggs from one cow can be implanted in the womb of another).
109. Gaylin, *supra* note 107, at 48.
     See also R. Scott, *The Body as Property*, Ch. 8 (1981).
110. *Id.*
111. D. Rorvik, *Brave New Baby* 109 (1971).
112. G. Taylor, *The Biological Time Bomb* 23–25 (1968).
113. G. Leach,*The Biocrats* 94 (1970).
114. J. Watson, Potential Consequences of Experimentation with Human Eggs, Jan. 28, 1971 (Papers 1, 3, 4, Harv. Univ. Biological Labs).
115. Lederberg, Experimental Genetics and Human Evolution, 100 AM. *Naturalist* 549, 562 (1966); Watson, Moving Toward the Clonal Man, *Atlantic Monthly* 50, 51 (May, 1971).
116. Comment, Asexual Reproduction and Genetic Engineering: A Constitutional Assessment of the Technology of Cloning, 47 *S. Cal. L. Rev.* 476 (1974).
117. *Supra* note 112, at 29.
118. *Id.* at 30.
119. *Supra* note 111, at 95.
120. *Id.* at 94.
121. Lederberg, Genetic Engineering or the Amelioration of Genetic Defect, 34 *Pharos* 9, 12 (1971).
122. *Id.* at 12.
123. Fletcher, Ethical Aspects of Genetic Controls, 285 *New Eng. J. Med.* 776, 779 (1971).
124. *Id.*
125. *Supra* note 116, at 561.
126. *Id.* at 550, 556.
127. Skinner v. Oklahoma, 316 U.S. 535, 541 (1942).
128. *Supra* note 115, at 550–552.
129. *Id.* at 556.
     See Shapiro v. Thompson, 399 U.S. 618, 638, n. 20.
130. *Supra* note 127, at 581–582; *U.S. Const.*, Art. I. 9, cl. 8; Amend, XIII.
131. *Supra* note 127, at 556.
132. *Id.* at 579.
     See R. Blank, *The Political Implications of Human Genetic Technology* 93–109, 117–122 (1983).
133. Vukowich, The Dawning of the Brave New World—Legal, Ethical and Social Issues of Eugenics, 1971 *U. Ill. L. F.* 189, 222.

If the challenged legislation incorporated negative, rather than positive, eugenic concepts so that it only restricted carriers of recessive debilitating defects from cloning, the constitutional problems would be minimized. The legitimacy of the state interest could not be challenged on the ground that it creates an elite group and therefore violates the nobility clause of the United States Constitution. A court could find readily that such a statute is rationally related to a legitimate state interest—specifically, diminishing the propagation of inferior traits. Scientific evidence more readily can provide a rational basis for the classification of those carrying debilitating defects than for those possessing superior genetic traits. Whether the state's interest in a negative eugenics program is sufficiently compelling to sustain the validity of the statute under a strict scrutiny test, however, is undertain. *Id.* at 198–201, 208.

134. Wilson, Foreword, *The Sociobiology Debate* at xi (A. Caplan ed. 1978).
135. E. Wilson, *Sociobiology: The New Synthesis* 578 (1975).
136. *Id.*
137. Peterson, Sociobiology and Ideas Become Real: Case Study and Assessment, 4 J. *Social Biol. Struct.* 125 (1981).
138. *Supra* note 135, at 575.
139. *Id.* at xiii, xiv.
140. Gould, Biological Potential vs. Biological Determinism in *The Sociobiology Debate* 343 (A. Caplan ed. 1978).
141. Dobzhansky, Anthropology and the Natural Sciences—The Problem of Human Evolution, 4 *Current Antrhopology* 146 (No. 138, 1963).
142. K. Boulding, *Ecodynamics: A New Theory of Social Evolution* (1978).
143. Caplan, Introduction, *The Sociobiology Debate* at 5 (A. Caplan ed. 1978).
144. Boulding, Sociobiology or Biosociology? *Society* 28 (Sept.–Oct. 1978).
    See also P. Singer, *The Expanding Circle* 27, 28 (1981); Frankel, Sociobiology and its Critics, *Commentary* 39 (July 1979).
145. Sociobiology Study Group of Science for the People, Sociobiology—Another Biological Determinism, in *The Sociobiology Debate* 280 at 287 (A. Caplan ed. 1978).
146. *Supra* note 143, at 3.
    See Gustafson, Sociobiology: A Secular Theology, *Hastings Center Rep.* 44 (Feb. 1979).

147. Singer, *supra* note 144, at 11.
148. *Id.* at 12.
149. *Id.*
150. *Id.* at 128.
151. *Id.* at 11.
152. J. Beckstrom, *Sociobiology and the Law* 13 (1985). *See also* D. Barash, *Sociobiology and Behavior* (1977).
153. Elliott, The Evolutionary Tradition in Jurisprudence, 85 *Colum. L. Rev.* 38 (1985).
154. *Id.* at 71.
155. O. Holmes, *The Common Law* 5 (M. Howe ed. 1963).
156. *Id.* at 32. See Holmes, Law in Science and Science in Law, 12 *Harv. L. Rev.* 443 (1899).
157. Beckstrom, Behavioral Research in Aid-Giving That Can Assist Lawmakers While Testing Scientific Theory, 1 *J. Contemp. Health L. & Pol'y* 25 (1985).
158. Beckstrom, Sociobiology and Intestate Wealth Transfers, 76 *Nw. U. L. Rev.* 216 (1981).
159. Rodgers, Bringing People Back: Toward A. Comprehensive Theory of Taking in Natural Resources Law, 10 *Ecology L.Q.* 205 (1982).
160. Hirshleifer, Privacy: Its Origins, Functions and Future, 9 *J. Legal Stud.* 649 (1980).
161. Rodgers, *supra* note 159, *Ecology* L.Q. at 219.
162. *Id.* at 221.
163. *See,* e.g., Epstein, A Taste for Privacy: Evolution and the Emergence of a Naturalistic Ethic, 9 *J. Legal Stud.* 665, 670 (1980).
164. *See* Nossal, "The Impact of Genetic Engineering on Modern Medicine," *Quadrant,* (Nov. 1983).
165. McGarity & Bayer, Federal Regulation of Emerging Genetic Technologies, 36 *Vand. L. Rev.* 461 (1983).
166. Smith, Quality of Life, Sanctity of Creation: Palliative or Apotheosis?, 63 *Neb. L. Rev.* 707 (1984).
167. G. Smith, *Genetics, Ethics and the Law* 164, 165 (1981).

CHAPTER 5

1. Seton, Mrs. Thatcher Volunteers for Transplant Donor Register, *The Times* (London), June 11, 1987, at 2, col. 2.

2. *Id.*
3. Childress, Some Moral Connections Between Organ Procurement and Organ Distribution, 3 *J. Contemp. Health L. & Pol'y* 85, 88 (1987).
4. *Id.* at 85.
5. Comment, The Law of Human Organ Procurement: A Modest Proposal, 1 *J. Contemp. Health L. & Pol'y* 195, 196 (1985).
6. Childress, *supra* note 3, at 87, 88.
7. *Id.* at 86. *See* National Organ Transplant Act, *Pub. L.* 98-507, 98 Stat. 2339–2348; 42 U.S.C. § 273–274 (1984).
8. *Id.*
9. *Id.*
10. *See* 8A *Unif. Laws Ann.* 15–16 (1983).
11. *Supra* note 3, at 88.
12. *Id.* at 89.
13. *Id.*
14. O'Rourke & Boyle, Presumed Consent for Organ Donation, *America* 326 (Nov. 22, 1986).
15. *Id.* at 327.
16. M. Shapiro, R. Spece, Jr., *Cases, Materials & Problems on Bioethics and Law* 740 (1981).
17. *Id.*
18. *Id.*
19. *Id.* at 767.
20. *Id.*
   Interestingly, the average recipient of a cadaver kidney has but a 20% chance of obtaining a transplant in the first year after the decision to transplant has been made. *Id.* at 769
   See generally Smith, Death Be Not Proud: Medical, Ethical and Legal Dilemmas In Resource Allocation, 3 *J. Contemp. Health L. & Pol'y* 47 (1987).
21. Dukeminier & Sanders, Organ Transplantation: A Proposal for Routine Salvaging of Cadaver Organs, 279 *New Eng. J. Med.* 413 (1968).
22. *Id.*
23. *Id.*
24. *Id.*
25. *Id.*
26. A commentator has suggested that persons should be required to donate tissues or organs if: "(A) the plaintiff shows by clear and convincing evidence: (1) that he is in *imminent* danger of dy-

ing from a disease that can be treated by transplantation . . . ; (2) that he stands to experience *substantial benefit* from such a transplant with the defendant serving as donor; (3) that transplantation from the defendant is the *exclusive* mode of treatment that offers the prospect of *substantial benefit* to the plaintiff; and (4) that the organ, tissue, or fluid sought is *expendable* by the donor—given the quantity of tissue or fluid to be removed and its regenerative capacity—and that the removal of the organ, tissue, or fluid will not result in disfigurement;" and "(5) the court finds that the risks to the defendant/donor are greatly outweighed by the benefits to the plaintiff/patient." Comment, Coerced Donation of Body Tissues: Can We Live with McFall v. Shimp?, 40 *Ohio St. L. J.* 409, 415–416, 420–421 (1979).

27. Thompson, The Eerie World of Living Heads, *Wash. Post*, Feb. 14, 1988, at C3, col. 1.

28. *Id.*

29. *Id.*

30. *Id.*

31. Salamone, The Problems of Neomorts, *Wash. Post Health Mag.*, Nov. 11, 1986, at 16.

32. *Id.*

33. *Id.*

34. Dukeminier, Supplying Organs for Transplantation, 68 *Mich. L. Rev.* 811 (1970).

35. *Id.*

36. *Id.*

37. Caplan, Should Fetuses or Infants be Utilized as Organ Donors?, 1 *Bioethics* 1 (1987).

   See Mahowald, Silver & Ratcheson, The Ethical Options in Transplanting Fetal Tissue, *Hastings Center Rep.* 9 (Feb. 1982); Hubbard, The Baby Fae Case, 6 *Med. & Law* 385 (1987).

38. Surgeons to Pursue Animal-to-Human Transplants, *Wash. Post Health Mag.*, Mar. 10, 1987, at 20.

39. *Id.*

40. Caplan, *supra* note 37.

41. *Id.* at 4.

   See Capron, Anencephalic Donors: Separate the Dead from the Dying, *Hastings Center Rep.* 5 (Feb. 1987).

42. Caplan, *supra* note 37, at 5.

43. *Id.* at 6.

See Smith, Intimations of Life: Extracorporeality and The Law, 21 *Gonz. L. Rev.* 395 (1986).

The National Institute of Health announced that it is investigating whether doctors are taking tissue from live fetuses for use in research before they are declared dead. This action was undertaken in response to a charge by the Foundation on Economic Trends and social activist Jeremy Rivkin that "physicians do not perform tests on aborted fetuses to determine whether they are dead prior to removing tissues or organs" for research use. Hilts, NIH Probes Allegations of Live Fetal Tissue Use, *Wash. Post*, Sept. 9, 1987, at A4, col. 1.

Prompted by an NIH proposal to implant fetal tissue into patients with Parkinson's disease, the Reagan Administration halted all federal research in the field. Specter, NIH Told to Stop Use of Aborted Fetuses: Controversial, Promising Tests Banned by Administration, *Wash. Post*, April 15, 1988, at 1, col. 4.

An NIH panel made preliminary findings, however, that using tissue from aborted fetuses for research and transplantation was acceptable. A final report to NIH on the issue will be presented in December, 1988, and draft guidelines submitted. Specter, Fetal Tissue "Acceptable" For Research: NIH Panel Avoids Stance on Abortion, *Wash. Post*, Sept. 17, 1988, at 1, col. 4. The White House is proposing an executive order that would ban all research conducted or funded by the federal government using fetal tissues obtained from induced abortions, and allow research only when the fetus died spontaneously. Rich & Spector, Bowen Delays Decision on Fetal Tissue, *Wash. Post*, Sept. 16, 1988, at A3, col. 4.

44. Caplan, *supra* note 37, at 7. *See* Gianelli, Anencephalic Heart Donor Creates New Ethics Debate, *Am. Med. News* at 51 (Nov. 6, 1987).

See Trafford, Fetal Tissue and Fine Lines: New Technologies Test the Power of Utilitarian Principles, *Wash. Post Health Mag.*, at 7, col. 1, Sept. 20, 1988, where the conclusion is made that it is impossible to separate the medical debate over fetal research from the political debate over abortion.

45. R. Fox, J. Swazey, *The Courage to Fail* 33–34 (1974).

46. Kirby, Bioethical Decisions and Opportunity Costs, 2 *J. Contemp. Health L. & Pol'y* 7, 21 (1986).

47. *Id.* at 20.

48. Note, Scarce Medical Resources, 69 *Colum. L. Rev.* 620, 653 (1969).

49. Gillie, Cuts at Guys: Now Hit Kidney Victims, *The Sunday Times* (London), Mar. 3, 1985, at 4, col. 4.
50. Smith, *Triage*: Endgame Realities, 1 *J. Contemp. Health L. & Pol'y* 143 (1985).
51. Childress, *supra* note 3, at 98.
52. *Id.*
53. *Id.* at 104.
54. Battiata, Kidney Transplants Draw Foreigners Here, *Wash. Post,* Sept. 19, 1983, at 1, col. 2.
55. *Id.*
56. *Id.*
57. *See* Boy Awaiting 4th Liver Develops Pneumonia, *N.Y. Times,* April 30, 1987, at Y13, col. 1.
58. *Id.*
59. Evans, 25-Hour Transplant: Weary Parents Await Va. Girl's Recovery, *Wash. Post,* Mar. 5, 1987, at C1, col. 2.
60. *Id.*
61. *Id.*
62. *Supra* note 16, at 822.
    See generally Curran, A Problem of Consent: Kidney Transplantation in Minors, 34 *N.Y.U. L. Rev.* 891 (1959).
63. *Supra* note 16, at 768.
    The State Attorney General of Florida brought an investigation recently to probe charges that employes of an organ procurement agency—The Florida Regional Bone and Tissue Bank and The Florida West Coast Organ Procurement Foundation—secretly harvested human body parts and sold them to overseas buyers. *Wash. Post,* Mar. 11, 1987, at A7, col. 1.
    See also Engel, Va. Doctors Plans Company to Arrange Sale of Human Kidneys, *Wash. Post,* Sept. 19, 1983, at A9, col. 1, where plans to open a company that would broker human kidneys for sale by arranging for donors throughout the world—and especially Third World countries—to sell one of their kidneys (for $10,000) were disclosed.
64. Smith, *Triage*: Endgame Realities, 1 *J. Contemp. Health L. & Pol'y* 143, 145 (1985).
    See generally Sanders & Dukeminier, Medical Advance and Legal Lag: Hemodialysis and Kidney Transplant, 15 *U.C.L.A. L. Rev.* 357 (1968); Smith, Death Be Not Proud: Medical, Ethical and Legal Dilemmas in Resource Allocation, 3 *J. Contemp. Health L. & Pol'y* 47 (1987).

65. *Id.* at 146.
66. *Id.*
  See generally Pellegrino, Rationing Health Care: The Ethics of Gatekeeping, 2 *J. Contemp. Health L. & Pol'y* (1986).
67. Smith, *supra* note 64, at 147.
  Confronted with sparse health funds, the state of Oregon decided that it could no longer pay for most organ transplants for its citizens. The money that would have been allocated for this effort can now be utilized for the prenatal care of some 1,500 pregnant women. As the Governor of the state commented: "How can we spend every nickel in support of a few people"— about thirty organ transplants costing as much as $100,000.00 per case—"when thousands never see a doctor or eat a decent meal?" Specter, Rising Costs of Medical Treatment Forces Organ to 'Play God', *Wash. Post*, Feb. 5, 1988, at 1, col. 1.
68. *Id.*
69. *Id.*
70. *Id.*at 148.
  See G. Smith, *Genetics, Ethics and the Law* 2, 8 (1981).
71. *Id.* at 148.
72. *Id.*
73. *Id.*
74. *Id.* at 149.
75. *Id.*
  See Freud, Organ Transplants: Ethical and Legal Problems in *Moral Problems in Medicine* at 44 (S. Gorowitz et al. eds. 1976).

CHAPTER 6

1. McShane, Hopkins Saves life with New Techniques, *Wash. Post*, Oct. 12, 1983, at C1, col. 5.
2. *Time*, June 22, 1981, at 71.
3. *Newsweek*, July 7, 1980, at 8.
  Because of these expenses, even estimated as high as $125,000.00 in order to assure the care and maintenance of the cryon until re-animated in the 21st century, one cryotorium, notably the Institute for Cryobiological Extension (ICE) in Los Angeles, has chosen to preserve cryonically on "ice" just the head of the cryon for later grafting onto new bodies. Thompson, The

Eerie World of Living Heads, *Wash. Post*, Feb. 14, 1988, at C3, col. 1.

4. *Newsweek*, August 16, 1976, at 8.

5. R. Prehoda, *Suspended Animation: The Research Possibility That May Allow Man to Conquer the Limiting Chains of Time* 10 (1969).

6. *Id.* at 11.

7. *Id.* at 7.

8. *Id.*
   See Klebanoff & Phillips, Temporary Suspensions of Animation Using Total Body Perfusion and Hypothermia: A Preliminary Report, 6 CRYOBIOLOGY 121–125, (Sept.–Oct., 1969).

9. *Supra* note 5, at 9.

10. A. Smith, *Current Trends in Cryobiology* (1970).

11. R. Ettinger, *Man into Superman* 251 (1972).

12. *Supra* note 5, at 73.

13. *Id.* at 13.

14. Mazur, Cryobiology: The Freezing of Biological Systems, 168 *Science* 939–949 (1970).

15. Dukeminier & Sanders, Organ Transplantation: A Proposal for Routine Salvaging of Cadaver Organs, 279 *New Eng. J. Med.* 413 (1969).

16. B. Luyet & M. Gehenio, *Life and Death at Low Temperatures* (1940).

17. R. Nelson, We *Froze the First Man* (1968).

18. *Id.* at 48.

19. *Id.*

20. J. Huxley, *Science, Religion and Human Nature* (1930).

21. Albano, *The Medical Examiner's Viewpoint in the Moment of Death* 19 (A. Winter ed., 1969).

22. J. Haldane, *Daedalus: Or Science and the Future* (1924).

23. J. Tuccille, *Here Comes Immortality* (1973); L. Kavaler, *Freezing Point: Cold as a Matter of Life and Death* (1970); R. Ettinger, *The Prospect of Immortality* (1964).

24. A. Harrington, *The Immortalist: An Approach to the Engineering of Man's Divinity* (1969).

25. *Id.* at 20.

26. *Id.* at 61.

27. *Id.* at 241.

28. Bryant & Snizek, The Iceman Cometh: The Cryonics Movement and Frozen Immortality, 11 *Society* 58 (Nov.–Dec. 1973).

29. R. Ettinger, *Man into Superman* (1972).

30. *Supra* note 24.

31. *Supra* note 29, at 215.
32. Kavaler, *supra* note 23, at 258.
33. *Supra* note 28, at 56, 60, 61.
34. Kavaler, *supra* note 23, at 228.
35. In 1976, it was estimated that the cost of preparation and indefinite storage was approximately $50,000. *Newsweek*, Aug. 16, 1976, at 11.
    See *supra* note 3 where the current estimated cost runs as high as $125,000.
36. *Time*, June 22, 1981 at 77.
37. *Supra* note 29.
38. E. Rievman, *The Cryonics Society: A Study of Variant Behavior Among Immortalists* 92 (1976).
39. Burger, Reflections on Law and Experimental Medicine in 1 *Ethical, Legal and Social Challenges to a Brave New World* at 211 (G. Smith, ed. 1982).
40. M. Shapiro & R. Spece, Jr., *Cases, Materials and Problems on Bioethics and Law* (1981).
41. Task Force on Death and Dying of The Institute of Society, Ethics and Life Sciences, Refinements in Criteria for Determination of Death, 221 *J.A.M.A.* 48 (1972).
42. *Id.*
43. Jeddelah, The Uniform Anatomical Gift Act and a Statutory Definition of Death, 8 *Transplantation Proceedings* No. 1, at 245 (1976).
44. Dukeminier, Supplying Organs for Transplantation, 68 *Mich. L. Rev.* 811 (1970).
45. International Comments, Declaration of Sydney, 206 *J.A.M.A.* 657 (1968).
46. Minutes of the Eleventh Meeting of The President's Commission for The Study of Ethical Problems in Medical and Biomedical and Behavioral Research, Washington, D.C., at 3, July 9, 1983.
    If cryonic suspension were to be recognized as an heroic measure designed to sustain life, then the concept of "mercy freezing" might have some legal validity and would thus be defined simply as freezing a terminally ill patient before clinical death occurs. R. Ettinger, *Man into Superman* (1972).
47. R. Ettinger, *The Prospect of Immortality* 3 (1964).
48. J. Gray, *The Rule Against Perpetuities* (1942).
49. Schuyler, The New Biology and The Rule Against Perpetuities, 15 *U.C.L.A. L. Rev.* 420 (1968).
50. G. Smith, *Medical-Legal Aspects of Cryonics: Prospects for Immortality* 15–23 (1983).

51. Smith, Uncertainties on the Spiral Staircase: Metaethics and The New Biology, 41 *The Pharos Med. J.* 10 (1978).
    See generally, Smith, Manipulating the Genetic Code: Jurisprudential Conundrums, 64 *Geo. L. J.* 697 (1976).

CHAPTER 7

1. *See* R. Gottfried, *The Black Death: Natural and Human Disaster in Medieval Europe* (1983).
2. *See* O'Brien, Aids and The Family, in *Aids and the Law*, Ch. 6, (W. Dornett ed. 1987).
    See generally, B. Tuchman, *A Distant Mirror: The Calamitous 14th Century* (1978).
3. Morgenstern, The Role of The Federal Government in Protecting Citizens from Communicable Disease, 47 *Cin. L. Rev.* 537, 541 (1978).
4. *Id.* at 539.
    See Comment, Protecting the Public from AIDS: A New Challenge to Traditional Forms of Epidemic Control, 2 *J. Contemp. Health L. & Pol'y* 191 (1986).
5. Specter, Report on AIDS: Trend Discouraging, *Wash. Post*, April 17, 1987, at A3, col. 4.
6. Surgeon General Koop: The Right, The Left and The Center of The AIDS Storm, *Wash. Post Health Mag.*, Mar. 24, 1987, at 6.
7. *World Press Rev.*, Sept. 1987, at 51.
8. O'Brien, *supra* note 2.
9. *Id.*
10. *Id.*
11. *Id.*
12. *Supra* note 6.
13. O'Brien, *supra* note 2.
14. *Id.*
15. *Id.*
    See generally Report of the Presidential Commission on The Human Immunodeficiency Virus Epidemic (June 24, 1988).
16. Okie, AIDS Tool Masks Immensity of Threat, *Wash. Post*, Mar. 26, 1987, at 1, col. 3.
17. *Id.*
    But see Boodman, Virus Moving Beyond Local Risk Groups:

Health Officials Say Heterosexuals, Especially Men, Have Not Changed Habits, *Wash. Post*, Mar. 27, 1987, at 10, col. 1.

18. Thompson, Debate Surfaces Over an AIDS Treatment, *Int'l Herald Tribune*, June 6–7, 1987, at 3, col. 5.

19. *Id.*

20. Buckley, What Can We Do About AIDS, *Wash. Post*, April 23, 1987, at A23, col. 6.

21. Carey & Arthur, The Developing Law on AIDS in the Workplace, 45 *Md. L. Rev.* 284, 287 (1987).

22. *Id.*

It is estimated that an individual exposed to the virus has a 20% risk of developing "full-blown AIDS" and a 25% risk of developing a related condition. *Id.* n. 11, at 287.

23. *Id.*

24. *Id.*

25. *Id.* at 288.

See AIDS: *Information/education Plan to Prevent and Control AIDS in the United States*, U.S. Dept. of Health & Human Services (Mar. 1987). In issuing this pamphlet-brochure, the Secretary of Health and Human Services cautioned that the scope and content of this AIDS education effort should be determined locally and be consistent with parental values. *Id.*, Preface.

Intravenous drug users now account for 17 percent of the nation's drug cases and are widely regarded as the most likely pathway for the disease to spread to the heterosexual population. *Newsweek*, April 13, 1987, at 63.

26. *Supra* note 21, at 288.

27. *Id.*

28. Comment, Protecting the Public from AIDS: A New Challenge to Traditional Forms of Epidemic Control, 2 *J. Contemp. Health L. & Pol'y* 191, 209–214 (1986).

See generally Comment, The Rights of an AIDS Victim in Wisconsin, 70 *Marq. L. Rev.* 55 (1986).

29. 480 U.S. 273, 55 U.S.L.W. 4245 (Mar. 3, 1987); 94 L. Ed. 2d 307; 107 S. Ct. 1123 (1987).

30. Reprinted in Daily Labor Rep. (BNA) No. 122, at D-1 (June 25, 1986).

31. *Supra* note 21, at 290.

32. *Id.*

33. 612 F. 2d 644 (2d Cir. 1979).

See generally AIDS in the Classroom: Room for Reason Amidst Paranoia, 91 *Dick. L. Rev.* 1055 (1987).

34. *Supra* note 29.
35. Carey & Arthur, The Developing Law on AIDS in the Workplace, 46 *Md. L. Rev.* 284, 291 (1987).
36. Goldberg, The Meaning of Handicapped, 73 *A.B.A.J.* 56 (1987).
37. *Supra* note 34.
    See Stewart, Good News for AIDS Victims, Refugees, 73 *A.B.A.J.* 50 (May 1, 1987).
38. 772 F. 2d 759, 764 (11th Cir. 1985).
39. *Supra* note 29, 55 U.S.L.W. at 4250.
40. *Id.* at 4248.
41. Stewart, *supra* note 37.
42. Kamen, AIDS Rules A Protected Handicap: Court Says Coverage of Bias Law Extends to Contagious Ill, *Wash. Post*, Mar. 4, 1987, at 1, col. 1.
43. *Id.*
    See generally Comment, Opening the Schoolhouse for Children with AIDS: The Education for all Handicapped Children Act, 13 *B. C. Envt'l Affairs L. Rev.* 583 (1987).

    There are also difficult problems for persons in the military services who contract AIDS. *See* Howe, Ethical Problems in Treating Military Patients with Human Immunodeficiency Virus Diseases, 3 *J. Contemp. Health L. & Pol'y* 111 (1987).

    The Department of Defense, two years after launching the world's most extensive AIDS screening program testing nearly 4 million people, has identified 5,890 carrying the deadly virus. Black, AIDS in Military: 5,890 Positive of 3.9 Million Tested, *Wash. Post*, Feb. 11, 1988, at A25, col. 1.
44. *Id.*
    Interestingly, under the Federal Occupational Safety and Health Act of 1970 (OSHA), 29 U.S.C. §§ 651–678 (1982), an employer must furnish employment in a workplace "free from recognized hazards that are causing or are likely to cause death or serious physical harm. . . ." (Sec. 654(a) (1)). OSHA also prohibits employers from retaliating against employees who refuse to be exposed to a health hazard that they have asked the employer to correct and that they in good faith reasonably believe poses a danger of death or serious injury. (Sec. 660(c)(1). Currently, no federal occupational safety and health standard exists for dealing specifically with AIDS in the workplace. Care & Arthur, *supra* note 21, at 317.

45. Sullivan, U.S. Insurance Co's Planning to Link Coverage to an AIDS Test, *Int'l Herald Tribune*, June 6–7, 1987, at 1, col. 3.

   See Commentaries on AIDS and Insurance, 100 *Harv. L. Rev.* 1782 *passim* (1987).

46. Cimons, Corporate Executives Urge AIDS Policy for Workplace, L.A. Times, Jan. 21, 1988, at 1, col. 5.

47. *Id.*

48. Ellsworth-Jones, Hopeful Signs from New AIDS Figures, *The Sunday Times* (London), July 5, 1987, at 12, col. 2, quoting the conservative columnist Norman Podhertz.

   See Nicholas, AIDS—A New Reason to Regulate Homosexuality, 11 *J. Contemp. L.* 315 (1984).

49. White, Mormon Links AIDS to 'Sexual Adventurism," *Wash. Post*, April 5, 1987, at 5, col. 1.

50. The United States Supreme Court held in *Bowers v. Hardwick*, 106 S. Ct. 2841, 2843 (1986) that no fundamental right existed for homosexuals to engage in sodomy.

51. Surgeon General Koop: The Right, The Left and The Center of the AIDS Storm, *Wash. Post Health Mag.*, Mar. 24, 1987, at 6.

   See Okie, Majority See AIDS as Public Threat, *Wash. Post*, Mar. 12, 1987, at A4, col. 1.

   See also Anderson & Spear, Explicit Anti-AIDS Campaign Debated, *Wash. Post*, Nov. 21, 1985, at 11, col. 6.

52. Hsu, Sex Ethics Curriculum Discussed: Va. Schools Must Teach Chastity, Parents Say, *Wash. Post*, April 28, 1987, at 5, col. 1.

53. *See* the Plan to Prevent and Control AIDS at 11.

54. Hilts & Boodman, Plan Issued for AIDS Education, *Wash. Post*, Mar. 17, 1987, at 1, col. 5.

55. Findley, When It Comes to AIDS, Many Teens Are Ignorant, *USA Today*, May 5, 1987, at D1, col. 3.

56. Carlin, Not By Condoms Alone, *Commonweal* 137 (Mar. 13, 1987).

57. *Supra* note 51, at 7.

58. *Id.*

59. *Id.*

60. *Id.* at 10.

61. *Id.*

   See Editorial, AIDS Education, *Wash. Post*, Feb. 28, 1987, at A22, col. 1.

62. Eliason, U.S. AIDS Brochure to go to All Households, *Wash. Post*, Jan. 28, 1988, at A8, col. 1.

63. *Supra* note 56.
64. Hodgkinson, Doctors Attack Cards to Prove AIDS-Free Status, *The Sunday Times* (London), July 5, 1987, at 4, col. 1.
65. *Id.*
66. Streitfeld, Putting Your AIDS Cards on The Table, *Wash. Post*, May 14, 1987, at C5, col. 2.
67. *Id.*
68. *Id.*
    See Comment, You Never Told Me . . . You Never Asked; Tort Liability for The Sexual Transmission of AIDS, 91 *Dick. L. Rev.* 529 (1986).
69. Wilentz, Putting AIDS to the Test, *Time*, Mar. 2, 1987, at 60.
70. *Id.*
    See Moore, Weinberger Approves AIDS Policy Overhaul, *Wash. Post*, April 23, 1987, at 1, col. 1.
    The Secretary of Defense approved new AIDS policy provisions that allow for the revocation of security clearances and a denial of access to classified information to military personnel who test positive for the AIDS virus. The policy also denies reservatists medical treatment for the disease at military hospitals and clinics. *Id.*
    See Feinberg, U.S. to Test Third of New Babies for AIDS: No Consent Sought, Blood Samples Won't Be Identified by Name, *Wash. Post*, Aug. 23, 1988, at 1, col. 1, Aug. 23, 1988.
    See also Howe, *supra* note 43.
71. Thompson, AIDS Blood Testing: What It Shows, *Wash. Post Mag.*, May 19, 1987, at 7.
72. *Id.*
73. Boodman, Koop: Mandatory Tests Would Harm Fight, *Wash. Post*, May 2, 1987, at A3, col. 1.
74. *Id.*
75. *Id.*
76. Cal. ex rel. Agnost v. Owen, No. 830321 (Cal. Super. Ct. San Francisco (1984) as reported in Comment, Protecting the Public from AIDS: A New Challenge to Traditional Forms of Epidemic Control, 2 *J. Contemp. Health L. & Pol'y* 191 at 203, n. 78 (1986).
77. *See* Laws to Combat AIDS Enacted in Illinois, *Wash. Post*, Sept. 22, 1987, at A3, col. 1; Johnson, Broad AIDS Laws Signed in Illinois, *N.Y. Times*, Sept. 22, 1987, at 7, col. 1; Baker & Schmidt, Va., Md. Lawmakers Search for Legislative Solutions to AIDS Epidemic, *Wash. Post*, Feb. 21, 1988, at C1, col. 1.

78. ILL. *Ann. Stat.* ch. 111 1/2, paras. 7401–7410 (Smith-Hurd 1988).
79. *Id.*
80. *Id.*
81. ILL. *Ann. Stat.* ch. 111 1/2, paras. 7301–7316 (Smith-Hurd 1988).
82. ILL. *Ann. Stat.* ch. 38, para. 1005-5-3; ch. 40, para. 204; ch. 111 1/2, para. 22.12a, 5606; ch. 126, para. 21 (Smith-Hurd 1988).

    Other states are following the Illinois legislature's lead in attempting to combat AIDS. *See*, e.g., *La. Rev. Stat.* §§ 9:241 & 242 (A); 34.3; 40:1099; 40:1299.131; 47:120.61; 241.242 (1987); the Communicable Disease Prevention and Control Act, *Tex. Rev. Civ. Stat. Ann.*, tit. 71, art. 4419a-1 (Vernon Supp. 1987).

    A 50-state survey is presented in *A Summary of AIDS Laws from the 1986 Legislative Sessions*, Geo. Wash. Univ., Intergovernmental Health Policy Project (Jan. 1987).
83. Pauling, Foreword, Symposium—Reflections on the New Biology, 15 U.C.L.A. L. Rev. 267, 270 (1986).
84. *New York Times*, Mar. 18, 1986, at A27, col. 4.
85. *See* Duncan, Public Policy and The AIDS Epidemic, 2 *J. Contemp. Health L. & Pol'y* 169 (1986).
86. *Int'l Herald Tribune*, July 1, 1987 at 1, col. 6.

    In Munich, Germany, a man who raped a woman with the full knowledge that he carried the AIDS virus, and could thereby transmit it to his victim, was found guilty by a West German Court of not only rape, but attempted murder. Man With AIDS Did Not Tell Partners, *Sydney Morning Herald*, July 22, 1987, at 12, col. 4. It was also proved that the man knew he was an AIDS carrier since 1985, but nonetheless continued to have sex with several women.

    The Soviet Union's strict anti-AIDS legislation, recently enacted, provides for a 5-year jail term for carriers of the lethal virus who have sexual contact with another person, even if the infection is not passed on. Lee, Soviets Pass Strict Law to Stem Spread of AIDS, *Wash. Post*, Aug. 26, 1987, at 1, col. 5.
87. Korematsu v. United States, 323 U.S. 214 (1944).
88. Duncan, *supra* note 73.
89. Heitz, AIDS Infection Found in 11 percent of U.S. Prostitutes, *Wash. Post*, May 27, 1987, at 1, col. 3.

    See Okie & Wheeler, D.C. Prostitutes Show a 50 Pct. AIDS Virus Rate, *Wash. Post*, Mar. 4, 1987, at 1, col. 1.
90. *Int'l Herald Tribune*, July 4–5, 1987, at 3, col. 4.
91. DeYoung, World Officials Meet to Exchange Date on AIDS, *Wash. Post*, Jan. 27, 1988, at A26, col. 1.

92. Marsh, AIDS Cases Forecast to Reach 1 M by 1991. *Fin. Times Ltd.*, Jan. 27, 1988, at 1, col. 1.
93. *Id.*
    Children comprise as many as one-third of all AIDS cases in some African countries and if this figure increases, it could wipe out the significant reductions in infant mortality achieved in recent decades. Hilts, AIDS Takes Heavy Toll of African Children, *Wash. Post*, Oct. 10, 1987, at A1, col. 1.
94. Hodgkinson, 1 in 10 AIDS Toll Looming over America, *The Sunday Times* (London), June 7, 1987, at 9, col. 6.
95. Buckley, What Can Be Done About AIDS, *Wash. Post*, April 23, 1987, at A23, col. 6.
    A related cost issue is tied to the matter of who should bear the costs of the still experimental anti-AIDS drug, AZT, once it is approved as a prescription drug and made available to large number of AIDS victims; for the cost will be anywhere from $7,000 to $15,000 for a year's supply. AZT is *not* a miracle drug for it does *not* defeat AIDS; it merely buys time for the afflicted patient, and, for that matter, adds to the overall costs of patient maintenance. Should the federal government underwrite totally a program that would entail support for *all* AIDS patients unable to purchase AZT? Manufacture and distribution of this drug could, arguably, under the basic principles of triage or salvageability, not be undertaken at all, simply because its users are *not* salvageable. Would such a policy be cruel and unusual punishment for the AIDS patients, or merely sound economics? One potential offset to AZT usage is taken to be a reduction in hospital stays and time lost from work. But, again, to what ultimate end? *See* Squires, The High Cost of Treating AIDS, *Wash. Post Health Mag.*, Mar. 10, 1987, at 7; Krauthammer, Saying No to AIDS Patients: Which Ones Get AZT, *Wash. Post*, Feb. 27, 1987, at A27, col. 6.
    See generally, McCormick, AIDS: The Shape of the Ethical Challenge, *America* 147 (Feb. 13, 1988).
96. Prentice, Jeers and Boos Greet Reagan AIDS Plan, *The Times* (London), June 2, 1987, at 20, col. 1. Indeed, it has been urged that the most effective policy to identify carriers of AIDS would be to screen the entire population for the disease. Duncan, Public Policy and The AIDS Epidemic, 2 *J. Contemp. Health L. & Pol'y* 169, 170 (1986). *See supra* note 43.
    Health and Human Services Secretary, Otis R. Bowen, told

a congressional panel yesterday that the Reagan administration opposes legislation to protect persons infected with the AIDS virus because "each state should be able to set its own rules." Boodman, Bill to Protect AIDS Carriers Opposed, *Wash. Post*, Sept. 22, 1987, at A3, col. 1.

97. Spector, AMA Committee Rejects U.S. Call for Broad Mandatory AIDS Testing, *Int'l Herald Tribune*, June 22, 1987, at 3, col. 4.

The Reagan Administration, as of December 1, 1987, required all persons seeking immigrant visas to the United States and all undocumented aliens seeking legalized status to undergo testing for the AIDS virus. Thornton, Immigrant AIDS Tests to Start, Dec. 1, *Wash. Post*, Aug. 29, 1987, at 1, col. 4.

98. Hardy, Rauch, Echenberg, Morgan & Curran, The Economic Impact of the First 10,000 Cases of AIDS in the United States, 255 *J.A.M.A.* 210 (1986).

99. *Supra* note 95.

100. *See* Carey & Arthur, The Developing Law on AIDS in the Workplace, 46 *Md. L. Rev.* 284 (1987).

101. *See generally* Smith, Manipulating the Genetic Code: Jurisprudential Conundrums, 64 *Geo. L. J.* 697 (1976).

See Orland, The AIDS Epidemic: A Constitutional Conundrum, 14 *Hofstra L. Rev.* 137 (1985).

102. Kirby, The Five Commandments for New Legislation on AIDS, Paper presented at The World Health Organization, Institut Merieux, Symposium on AIDS, Annecy, Switzerland, June 20–21, 1987.

See also Kirby, AIDS Legislation–Turning Up The Heat?, 60 *Australian L. J.* 324 (1986).

103. *Id.*

CHAPTER 8

1. B. Clark, *Whose Life Is It Anyway* 104 (1978).

2. *Id.* at 116.

3. *Id.* at 80.

4. Rational Suicide Raises Patient Rights Issues, 66 *A.B.A. J.* 1499 (Dec. 1980); *Time*, July 7, 1980, at 49.

5. *Time*, Mar. 14, 1983, at 96.

6. *Id.*
7. B. Rollin, *Last Wish* 149, 150, 175 (1985).
8. *Id.* at 187.
9. Bouvia v. County of Riverside, trial transcript reprinted in 1 *Issues L. & Med.* 485 (1986).
10. *Id.* at 491.
    See Annas, When Suicide Prevention Becomes Brutality: The case of Elizabeth Bouvia, *Hastings Center Rep.* 20 (1984); Van den Haag, The Bouvia Case: A Right to Die?, *Nat'l Rev.* 45 (May 4, 1984).
11. Bouvia v. High Desert Hospital, trial transcript reprinted in 1 *Issues L. & Med.* 493 (1986).
12. *Id.* at 498.
13. Bouvia v. High Desert Hospital, Slip Opinion.
14. *Id.*
    See generally Brown & Thompson, Nontreatment of Fever in Extended Care facilities, 300 *New Eng. J. Med.* 1246 (May 31, 1979).
15. 163 Cal. App. 2d 186 (1984).
16. *Id.* at 193.
17. *Id.*
18. *See* Schloendorff v. Soc. of N.Y. Hosp., 211 N.Y. 125, 105 N.E. 92, 93 (1914); Matter of Spring, 405 N.E. 2d 115 (Mass. 1980); Lane v. Candura, 376 N.E. 2d 1232 (Mass. App. 1978); Matter of Quackenbush, 383 A. 2d 785 (N.J. Morris County Ct. 1978); Matter of Conroy, 486 A. 2d 1209 (N.J. 1985); Satz v. Perlmutter, 362 So. 2d 160, 379 So. 2d 359 (1980); In re Osborne 294 A. 2d 372 (D.C. 1972); Supt. of Belchertown School v. Saikewicz, 370 N.E. 2d 417 (Mass. 1977).
19. *Supra* note 15.
20. 147 Cal. App. 3d 1006 (1983).
21. 379 So. 2d 359 (Fla. 1980).
    See Corbett v. D'Allessandro where it was held feeding tubes were *just* as any other medical care and, as such, to be used or withheld consistent with the wishes of the particular patient or on the proportion of benefits to burdens according to the standard appropriate under the facts of the case. It was held, in essence, that an extended right of privacy entitled one to refuse acts of artificial feeding even though a state law mandated that artificial forms of sustenance could not be withheld or withdrawn. 487 So. 2d 368 (Fla. Dist. Ct. App.) rev. denied (1986).

For similarly related cases on withholding of nutrition *see*: Brophy v. New England Sinai Hosp. Inc. 398 Mass. 417, 497 N.E. 2d 626 (1986); In re Jobes, 210 N.J. Super. 543, 510 A. 2d 133 (1986); In re Clark, 210 N.J. Super. 548, 510 A. 2d 136 (1986); in re Visbeck, 210 N.J. Super. 527, 510 A. 2d 125 (1986); In re Conroy, 98 N.J. 321 (1985); In the Matter of Hier, 18 Mass. App. Ct. 200, 464 N.E. 2d 959 (1984).

See also Paris & McCormick, The Catholic Tradition on the Use of Nutrition and Fluids, *America*, 356 (May 2, 1987).

22. M. Shapiro, R. Spece, Jr., *Cases, Materials & Problems on Bioethics and Law* 574 (1981).

23. Statement of the World Medical Assembly, 140 *Med. J. Aust.* 431 (Oct. 1983); President's Commission for The Study of Ethical Problems in Medicine and Biomedical and Behavioral Research, *Deciding to Forego Life-Sustaining Treatment* 88–89, 97, 134–136 (1983).

24. Pellegrino, Rationing Health Care: The Ethics of Medical Gatekeeping, 2 *J. Contemp. Health L. & Pol'y* 23 (1986); Smith, Triage: Endgame Realities, 1 *J. Contemp. Health L. & Pol'y* 143 (1985).

25. Smith, *supra*.

26. Kuhse, Euthanasia–Again?. 142 *Med. J. Aust.* 610 (May 27, 1985). *See* Sharma, Euthanasia in Australia, 2 *J. Contemp. Health L. & Pol'y* 131 (1986).

27. *Id.*

28. Kuhse, *supra* note 26, at 611.

29. *Id.*

30. Statement of the Council on Ethical and Judicial Affairs, American Medical Association, March 15, 1986, Chicago, Illinois.

31. Colburn, AMA Ethics Panel Revise Rules on Withholding Food, *Wash. Post Health Mag.* 9 (April 2, 1986).

32. *Supra* note 30.

33. *Supra* note 31.

34. Ginnex, A Prosecutions View on Criminal Liability for Withholding Medical Care: The Myth and The Reality, in *Legal and Ethical Aspects of Treating Critically and Terminally Ill Patients* (E. Doudera & W. Peters eds. 1982).

35. Newman, Treatment Refusals for the Critically and Terminally Ill: Proposed Rules for The Family, The Physicians and The State, 3 *N.Y. Law Sch. Human Rights Annual* 35 (1985).

36. Supt. of Belchertown State School v. Saikewicz, 373 Mass. 728, 743 n. 11, 370 N.E. 2d 417, 426 n. 11 (1977).

37. *Supra* note 35, at 76.
38. *Id.* at 79.
39. *Id.* at 44.
40. *Id.*
41. *Id.* at 46.
    See Smith, Quality of Life, Sanctity of Creation: Palliative or Apotheosis?, 63 *Neb. L. Rev.* 707 (1984).
42. *Id.* at 79.
43. Dagi, The Ethical Tribunal in Medicine in 1 *Ethical, Legal & Social Challenges to a Brave New World*, at Ch. 7 (G. Smith ed. 1982).
44. Ala. Code §§ 22-8A-1 to -10 (1984); Ariz. Rev. Stat. Ann. §§ 36-3201 to -3210 (1986); Ark. Stat. Ann. §§ 82-3801 to -3804 (Supp. 1985); Cal. Health & Safety Code §§ 7185–7195 (West Supp. 1985); Colo. Rev. Stat. §§ 15-18-101 to -113 (Supp. 1986); 1985 Conn. Acts 85-606 (Reg. Sess.); Del. Code Ann. tit. 16, §§ 2501–2508 (1983); D.C. Code Ann. §§ 6-2421 to -2430 (Supp. 1986); Fla. Stat. §§ 765.01–.15 (1986); Ga. Code Ann. §§ 32-31-1 to -12 (1985 & Supp. 1986); Idaho Code §§ 39-4501 to -4508 (1985 & Supp. 1986); Ill. Ann. Stat. ch. 110 1/2, para. 701–710 (Smith-Hurd Supp. 1986); Ind. Code Ann. §§ 16-8-11-1 to -21 (Burns Supp. 1986); Iowa Code Ann. §§ 144A.1 to .11 (West Supp. 1986); Kan. Stat. Ann. §§ 65-28-101 to -109 (1980); La. Rev. Stat. Ann. § 40:1299.58.1 to .10 (West Supp. 1986); Me. Rev. Stat. Ann. tit. 22, §§ 2921–2931 (Supp. 1986); Md. Health Gen. Code Ann. §§ 5-601 to -614 (Supp. 1986); Miss. Code Ann. §§ 41-41-101 to -121 (Supp. 1985); Mo. Ann. Stat. §§ 459.010 to .055 (Vernon Supp. 1986); Mont. Code Ann. §§ 50-9-101 to -111 (1985); Nev. Rev. Stat. §§ 449.540 -690 (1986); N.H. Rev. Stat. Ann. §§ 137-14:1 to -:16 (Supp. 1985); N.M. Stat. Ann. §§ 24-7-1 to -11 (1986); N.C. Gen. Stat. §§ 90-320 to -322 (1985); Okla. Stat. Ann. tit. 63, §§ 3101 to 3111 (West Supp. 1987); Or. Rev. Stat. §§ 97.050 to .090 (1985); Tenn. Code Ann. 32-11-101 to -110 (Supp. 1986); Tex. Rev. Civ. Stat. Ann. art. 4590h (Vernon Supp. 1986); Utah Code Ann. §§ 75-2-1101 to -1118 (Supp. 1986); Vt. Stat. Ann. tit. 18, §§ 5251 to 5262 (Supp. 1985); Va. Code Ann. §§ 54-325.8:1 to :13 (Supp. 1986); Wash. Rev. Code Ann. §§ 70.122.010 to .905 (Supp. 1986); W. Va. Code §§ 16-30-1 to -10 (1985); Wis. Stat. Ann. §§ 154.01 to .15 (West Supp. 1986); Wyo. Stat. §§ 33-26-144 to -152 (Supp. 1986).
45. President's Commission for The Study of Ethical Problems in Medicine and Biomedical and Behavioral Research, *Deciding to*

Forego Life-Sustaining Treatment: Ethical, Medical, and Legal Issues in Treatment Decisions 139 (1983).

46. *Id.* at 140.
47. *Id.*
48. *Id.*
49. *Id.* at 141.
50. *Id.*
51. Ala. Code §§ 28-8A-1 to 22-8A-10 (Supp. 1982); Ark. Stat. Ann. §§ 82-3801 to 82-3804 (Michie Supp. 1983); Cal. Health & Safety Code §§ 7185-95 (West Supp. 1983); Del. Code Ann. tit. 16, §§ 2501-09 (1983); D.C. Code Ann. §§ 6-2421 to 6-2430 (Michie Supp. 1983); Life-Prolonging Procedure Act of Florida, ch. 84-58, 3 Fla. Sess. Law Serv. 40 (West 1984); Ga. Code Ann. §§ 31-32-12 (Michie Supp. 1984); Idaho Code §§ 39-4501 to 39-4508 (Michie Supp. 1983); Ill. Ann. Stat. ch. 110 1/2, §§ 701–710 (Smith-Hurd Supp. 1984–85); Kan. Stat. Ann. §§ 65-28, 101 to 65-28, 101 to 65-28, 109 (1980)); Act. of Apr. 16, 1984, ch. 365, 1984 Miss. Law 98; Nev. Rev. Stat. §§ 449.540 to 449.690 (1983); N.M. Stat. Ann. §§ 24-77-1 to 24-7-11 (1981); N.C. Gen. Stat. §§ 90-320 to 90-323 (1981); Or. Rev. Stat. §§ 97.090 (1981); Tex. Rev. Civ. Stat. Ann. art. 4590h (Vernon Supp. 1983); Vt. Stat. Ann. tit. 18, §§ 5251–5262 (Supp. 1984); Va. Code §§ 54-325.8:1 to 54-325.8:13 (Michie Supp. 1984); Wash. Rev. Code Ann. §§ 70.122.010 to 70.122.905 (West Supp. 1983–84); W. Va. Code §§ 16-30-1 to 16-30-10 (Michie Supp. 1984); Wis. Stat. Ann. §§ 154.01 to 154.15 (Supp. 1985); Wyo. Stat. §§ 33-26-144 to 33-26-154 (Michie Supp. 1984).
52. Florida also recognizes a witnessed oral statement. Life-Prolonging Procedure Act of Florida, ch. 84-58, §§ 3–4, 3 Fla. Sess. Law Serv. 40, 41–42 (West 1984); see also Va. Code § 54-325.8:2 (Michie Supp. 1984).
53. Generally, the acts limit their coverage to life-sustaining procedures and do not cover curative measures. The definition of life-sustaining procedures varies among jurisdictions. Most acts define life-sustaining procedures as medical procedure or intervention which, by artificial or mechanical means, "only serves to prolong the moment of death." Excluded from the definition of life-sustaining procedures in most acts are procedures necessary to provide comfort, care, or to alleviate pain. *See*, e.g., Ill. Ann. Stat. ch. 1101/2 § 702 (Smith-Hurd Supp. 1984–85); Ore. Rev. Stat. § 97.050 (1981); Wyo. Stat. § 33-26-144 (Michie Supp. 1984).

54. Many acts define a terminal condition as an incurable condition resulting from illness, disease, or injury which would cause death regardless of the application of life-sustaining procedures. *See*, e.g., Cal. Health and Safety Code § 7186 (West Supp. 1983); N.M. Stat. Ann. § 24-7-2 (1981); Or. Rev. Stat. § 97.050 (1981); Tex. Rev. Civ. Stat. Ann. art. 459Oh § 2 (Vernon Supp. 1983); Vt. Stat. Ann. tit. 18, § 5252 (Supp. 1984); Wash. Rev. Code Ann. § 70.122.020 (West Supp. 1983–84); W. Va. Code § 16-30-2 (Michie Supp. 1984). The words "would cause death regardless of the application of life-sustaining procedures" seem to exclude the situation where a person may be kept alive on life support indefinitely. Other legislatures have avoided this problem by defining a terminal condition as a condition which would cause imminent death if life-support was not applied. *See*, e.g., Nev. Rev. Stat. § 449.590 (1983).

55. 12 Unif. Laws Ann. § 1 (1985 Supp.)

    See also Smith, Legal Recognition of Neocortical Death, 73 *Cornell L. Rev.* 850 (1986).

56. Humane and Dignified Death Initiative, Sec. 1(a) proposed as amendment to Art. 1 of the Calif. Const.

    See Parachini, The California Humane and Dignified Death Act Initiative, Hastings Center Rep. 10, (Jan.–Feb. 1989).

57. § 7188, Humane and Dignified Death Act.

58. *Id.* at § 7188.5.

59. *Id.* at § 7189(a).

60. *Id.* at § 7189(b), § 7195.

61. *Id.* at § 7191(c).

62. *Id.* at § 7192.

63. *Id.* at § 2443.

64. *Id.*

65. *See* Kuhse, The Case for Active Voluntary Euthanasia, 14 LAW, MED. & HEALTH CARE 145 (Sept. 1986); Engelhardt & Malloy, Suicide and Assisting Suicide: A Critique of Legal Sanctions, 36 Sw. U. L. J. 1003 (1982).

    See generally [Cardinal] Bernardin Condemns Legal Assisted Suicides: Congress Urged to Oppose the Practice, *Wash. Post*, Feb. 20, 1988, at C6, col. 1; Garbesi, The Law of Assisted Suicide, 3 *Issues L. & Med.* 93 (1987). *See* H. Con. Res. 194, 100th Cong., 1st Sess. (Oct. 7, 1987) expressing the sense of the Congress that efforts to allow people to assist others to commit suicide and

efforts to promote suicide as a rational solution to certain problems should be opposed.
66. C. Fried, *Right and Wrong* 146–147 (1978).
67. J. Mill, *On Liberty* 6 (1873). *See* G. Smith, *Final Choices*, (1989).

CHAPTER 9

1. *See* Smith, Artificial Insemination Redivivus, 2 *J. L. Med.* 113,
128–129 (1981).
2. Kritchevsky, The Unmarried Woman's Right to Artificial Insemination: A Call for an Expanded Definition of Family, 4
Harv. Women's L. J. 1, 26–39 (1981); Annas, Fathers Anonymous: Beyond the Best Interest of the Sperm Donor, 14 *Fam.
L. Q.* 1 (1980; Shaman, Legal Aspects of Artificial Insemination,
18 *J. Fam. L.* 330, 344–346 (1980).
3. *See* Smith, The Perils and Peregrinations of Surrogate Mothers,
1 *Int'l J. Med. & L.* 325 (1982); Comment, Surrogate Motherhood
in California: Legislative Proposals, 18 *San Diego L. Rev.* 341
(1981); Brophy, A Surrogate Mother Contract to Bear a Child,
20 *J. Fam. L.* 263 (1981).
4. *See The Artificial Family* (R. Snowden & G. Mitchell eds. 1981).
5. *See* Karst, The Freedom of Intimate Association, 89 *Yale L. J.*
624 (1980); Eichbaum, Sexual Expression, 10 *Colum. Human
Rights L. Rev.* 525 (1978–79); Wilkinson & White, Constitutional
Protection for Personal Life Styles, 62 *Cornell L. Rev.* 563 (1977).
6. A number of physical difficulties may impede a couple from
conceiving by normal sexual intercourse. The husband, for example, may suffer from retrogade ejaculation, physical impo-
tence, malformation of the penis, obesity, or low fertility. Likewise, impediments to conception in the wife may include
vaginal tumors or scarring, an abnormal position of the uterus,
a very small cervical opening, or obesity. *See also* Comment,
supra note 2, at 916–917.
    Most commentators agree that homologous insemination
poses few legal problems. *See* Note, Smith, Through a Test
Tube Darkly: Artificial Insemination and the Law, 67 *Mich. L.
Rev.* 127 (1968).
7. Artificial insemination by an unrelated and usually unidentified
donor is an alternative to childlessness when the husband is

absolutely or severely sterile. This may also be a preferred practice when the husband is the carrier of genetic defects or when an abnormal pregnancy is likely because of incompatible Rh blood factors. *See* G. Smith, *Genetics, Ethics and the Law* (1981).

    See Fitzgerald v. Ruckel where it was maintained that because of a physician's failure to properly genetically screen a donor for A.I.D., the possibility that the donor in fact carried a genetic anomaly which caused the issue subsequently born of the procedures to die of the syndrome, "failure to thrive," should impose liability. The Supreme Court of Nevada affirmed the lower court's determination that the physician was not negligent in his procedure and that there is no warranty of merchantability of fitness of a donor's sperm and, thus, no liability to the donor for a possible genetic defect being passed. Jan. 28, 1982, Sup. Ct. Order Dismissing Appeal, Slip Opinion in Case No. 11433. *See* Smith, Great Expectations or Convoluted Realities: Artificial Insemination in Flux, 3 *Fam. L. Rev.* 37 at 38 (1981).

8. Although some doctors disapprove of this practice, *see*, e.g., Holloway, Artificial Insemination: An Examination of the Legal Aspects 43 *A.B.A.J.* 1089, 1155–1156 (1957), it is used at times to give the husband some hope that he is in fact the natural father of the child. Shaman, *supra* note 2, at 332. For example, if the husband suffers from poor sperm motility (oligozoopermia), the combination of his sperm with that of a donor may result in the fertilization of the egg with his sperm. *See* Comment, Therapeutic Impregnation: Prognosis of a Lawyer—Diagnosis of a Legislature, 39 *Cin. L. Rev.* 201, 297 (1970).

9. Shaman, *supra* note 2, at 333.

    See also Fraser, Seven New Ways to Make a Baby, *The Weekend Australian*, April 7–8, 1984, at 16, col. 1.

10. *See* Smith, Through a Test Tube Darkly: Artificial Insemination and the Law, 67 *Mich. L. Rev.* 127, 130 (1968).

    But see Note, Artificial Insemination versus Adoption, 34 *Va. L. Rev* 822 (1948).

    The Catholic Church condemns AI of any kind both within and without marriage. *See* Address by Pope Pius XII, Fourth World Congress of Catholic Dcotors, Rome, Sept. 29, 1949, reproduced and discussed in G. Kelly, Medico-Moral Problems, 223–30 (1958). *See also* Lombard, Artificial Insemination Civil Law and Ecclesiastical Views, 2 *Suffolk L. Rev.* 137 (1968); Weisman, Symposium on Artificial Insemination—The Religious Viewpoints, 7 *Syracuse L. Rev.* 96, 99–106 (1955–1956).

In 1948, the Church of England held—consistent with the followers of the Lutheran faith, the Jewish Orthodox, and the Roman Catholic—that artificial insemination was not only wrong in principle but contrary to Christian beliefs. Today, because of widespread use of AID, the Church of England, as represented by its clergy, is no longer so rigid nor dogmatic in its condemnation of this process. R. Scott, *The Body as Property* 202 (1981).

11. Kritchevsky, *supra* note 2, at 1.
12. *See infra* note 20.
13. 152 N.J. Super. 160, 377 A.2d 821 (1977).

Initially, AID, even within the marriage, was not greeted with much enthusiasm by the judiciary. Some courts held that AID amounted to adultery, irrespective of the husband's consent, Orford v. Orford, 58 D.L.R. 251 (Ontario Sup. Ct. 1921); Doornbos v. Doornbos, 23 U.S.L.W. 2308 (unreported decision of Super. Ct., Cook County, Ill., Dec. 13, 1954), appeal dismissed on other grounds, 12 Ill. App. 2d 473, 139 N.E. 2d 844 (1956), and that a child so conceived was illegitimate. Gursky v. Gursky, 39 Misc. 2d 1083, 242 N.Y.S. 2d 406 (Sup. Ct. 1963).

The modern trend contradicts these earlier positions. The legal status of the AID child has been legitimized by both courts, In re Adoption of Anonymous, 74 Misc. 2d 99, 345 N.Y. 2d 430 (Sup. Ct. 1973); People v. Sorenson, 58 Cal. 2d 280, 437 P.2d 495, 66 Cal. Rptr. 7 (1958) and legislatures. *See infra* note 28. Moreover, most jurisdictions do not consider AID adultery since it does not encompass actual sexual intercourse. Shaman, *supra* note 2, at 334.

14. Smith, A Close Encounter of the First Kind: Artificial Insemination and An Enlightened Judiciary, 17 *J. Fam. L.* 41 (1978).
15. 152 N. J. Super. at 161, 377 A.2d at 822.
16. *Id.* at 164, 377 A.2d at 825.
17. *See* Kritchevsky, *supra* note 2, at 15 (arguing court could have employed an estoppel theory).
18. Turano, Paternity by Proxy, *Am. Mercury*, 419, 422 (April 1938).
19. Between 6000 to 10,000 children are born each year in the United States as a result of AID. Currie—Cohen, Luttrell, Leigh & Shapiro, Current Practice of Artificial Insemination by Donor in the United States, 300 *New Eng. J. Med.* 585, 588 (1979). More-

over, it is estimated that 250,000 people in the United States
have been conceived by artificial insemination. Kritchevsky,
supra note 2, at 1, n. 3, citing F. Mims & M. Swenson, *Sexuality:
A Nursing Perspective* 192 (1980).

The New England Journal of Medicine article pointed up
clearly the very real and serious concern about the potential for
incest, for it was found that sperm from one donor had been
used to produce 50 children! *Id.*

20. Alaska Stat. § 20.20.010 (1974); Ark. Stat. Ann. § 61–141(c)
(1981); Cal. Civ. Code Ann. § 7005 (West Supp. 1982); Cal.
Penal Code Ann. § 270 (West 1970); Colo. Rev. Stat. § 19-6-106
(1978); Conn. Gen. Stat. Ann. §§ 45-69 f-n, 45–152 (West Supp.
1980); Fla. Stat. Ann. § 742.11 (West Supp. 1981); Ga. Code
Ann. §§ 74-101.1, 74-9904 (1973); Kan. Stat. Ann. §§ 23-128 to
23-129 (1981), 23-130 (1981); La. Civ. Code Ann. art. 188 (West
Supp. 1982); Md. Est. & Trusts Code Ann. § 1-205(b) (1974);
Mich. Comp. Laws Ann. §§ 333.2824(6), 700.111(2) (1980);
Mont. Rev. Codes Ann. § 61-306 (Supp. 1977); Nev. Rev. Stat.
§ 126 061 (1979); N.Y. Dom. Rel. Law § 73 (McKinney 1977);
N.C. Gen. Stat. § 49-A-1 (1976); Okla. Stat. Ann. tit. 10 §§ 551-
553 (West Supp. 1980); Or. Rev. Stat. §§ 109.239(2), 109.243,
109.247, 677-355 (1979); Tenn. Code Ann. § 53-446 (Supp. 1981);
Tex. Fam. Code Ann. tit. § 12.03(a) (Vernon 1975); Va. Code
§ 64.1-7.1 (1980); Wash. Rev. Code Ann. § 26.26.050 (Supp.
1981–1982); Wis. Stat. Ann. § 767.47(9) (Special Pamphlet 1981),
§ 891.40(a) (West Supp. 1981–1982); Wyo. Stat. § 14-2-103
(1980).

21. For an analysis of the common elements found in the majority
of these statutes, *see* Comment, Artificial Human Reproduction:
Legal Problems Presented by the Test Tube Baby, 20 *Emory L.J.*
1045, 1065–1071 (1979).

See Smith, For Unto Us is Born a Child—Legally, 56
A.B.A.J. 143 (1970); Schuyler, The New Biology and the Rule
Against Perpetuities, 15 *U.C.L.A. L. Rev.* 420 (1968).

See generally Sappideen, Life After Death—Sperm Banks,
Wills and Perpetuities, 53 *Australian L. J.* 311 (June 1979).

22. Or. Rev. Stat. § 677.365 (1979) (emphasis added). One commentator has recently suggested that the statutes of California,
Colorado, Washington, and Wyoming, which follow the language of the Uniform Parentage Act but omit the word "married" from one of its provisions, may arguably legitimize the

practice for unmarried women. *See* Kritchevsky, *supra* note 2, at 18. In New South Wales, under the Artificial Conception Bill of 1984, given Royal Assent March 5, 1984, a child born as a result of artificial insemination by donor (AID) accomplished with the husband's consent is deemed to be the child of the husband. A child born from in vitro fertilization, where genetic material is provided by the husband and wife, or where semen is provided by a donor, is deemed the child of the husband and wife. The same principles are also applied to de facto relationships. Yet, the new law does not cover children born as a result of an in vitro procedure using donated eggs. The Equality of Status Amendment Bill, 1984, with respect to certain presumptions arising under that Act, was made cognate with The Artificial Conception Act of 1984.

23. One commentator seems to have interpreted the statute as limiting the practice to married women. *See* Shaman, *supra* note 2, at 344–345.

24. *See generally* Reitman v. Mulkey, 387 U.S. 369 (1967).

25. N. Keane, D. Breo, *The Surrogate Mother* (1981).

26. R. Blank, *The Political Implications of Human Genetic Technology* 68 (1981); G. Smith, *Genetics, Ethics and the Law* 110, 124, 125 (1981).

   See also Sappideen, the Surrogate Mother—A Growing Problem, 6 U. *New S. Wales L.J.* 79 (1983); Rassaby, Surrogate Motherhood: The Position and Problems of Substitutes in *Test-Tube Babies: A Guide to Moral Questions, Present Techniquest and Future Possibilities*, at 97 (W. Walters & P. Singer eds. 1982).

   The Senior Judge of The Family Court of Australia, Mr. Justice Asche, has summed up accurately two of the very real dangers of surrogation: duress and blackmail; duress, because the surrogate mother could very well decide to assume this status because of desperate poverty; and blackmail, because it is possible that a surrogate could endeavor to increase the original price for her services by threatening to keep the child upon its birth. Asche, Ethical Implications in the Use of Donor Sperm, Eggs and Embryos in The Treatment of Infertility, 57 [*Australian*] *Law Institute J.*, 716–719 (1983).

27. Keane, Legal Problems of Surrogate Motherhood, 1980 *S. Ill. U.L.J.* 147, 152.

   The theological response to this phenomenon is, as with artificial insemination, to regard it as an essentially adulterous act, since it is considered to be an "intrusion of a third party into the psycho-physical union of husband and wife." Kavanagh, Theologians Hit "Surrogate Mother" Business, *The Catholic Standard*, April 8, 1982, at 33, cols. 1–4.

28. *See*, e.g., Ky. Rev. Stat. § 199.590(2) (1980); Mich. Comp. Laws Ann. § 710.54 (Supp. 1983); *N.J. Stat. Ann.* § 9:3–54(a) (West Supp. 1983).

   In England, a special government inquiry into the complexities of the New Biology, the Warnock Commission, proposed an absolute ban on surrogate mothers. *Time*, Aug. 6, 1984, at 50.

   In Australia, The Report of the Committee to Consider the Social, Ethical and Legal Issues Arising from In Vitro Fertilization submitted in August, 1984, recommended "that surrogacy arrangements, shall in no circumstances be made at present as part of an IUF programme" in the State of Victoria. *Report on the Disposition of Embryos Produced by in Vitro Fertilization*, Sec. 4.17 (Aug. 1984). But see *infra* note 73, Chap. 11, 1988 Michigan Surrogate Parenting Act.

29. *See*, e.g., Adoption Hotline Inc. v. State of Florida 385 So.2d 682 (C.D.C.A. Fla. 1980); Comment, Independent Adoptions: Is the Black and White Beginning to Appear in the Controversy over Gray Market Allocations? 18 *Duq. L. Rev.* 629 (1980); Comment, Moppets on the Market: The Problem of Unregulated Adoptions, 59 *Yale L. J.* 715 (1950).

30. A. Corbin, *On Contracts* § 1792 (1962).

31. Erickson, Contracts to Bear A Child, in 1 *Ethical, Legal and Social Challenges to a Brave New World* 98, 100 (G. Smith ed. 1982).

32. Krucoff, Private Lives: The New Surrogate, *Wash. Post*, Sept. 24, 1980, at B5, col. 2.

   See generally Annas, Surrogate Embryo Transfer, *Hastings Center Rep.* 25 (June 1984).

33. Unif. Parentage Act of 1973, Sec 5(b) in *Unif. Laws Ann.*, Matrimonial, Family and Health Laws at 587 (1979). The Act, itself, has been adopted in but eight jurisdictions. *Id*.

34. Annas, Contracts to Bear a Child: Compassion and Commercialism, *Hastings Center Rep.* 23, 24 (April 1981).

35. *Id*.

   See generally Karst, The Freedom of Intimate Association, 89 *Yale L.J.* 624 (1980).

36. *Supra*, note 31; Harris, Stand-In Mother—Maryland Woman to Bear Child for Couple, *Wash. Post*, Feb. 11, 1980, at 1, col. 3.

37. *Id.*
38. *Id.*

   See generally Comment, Constitutional Limitations on State Intervention in Prenatal Care, 62 *Va. L. Rev.* 1051 (1981).
39. *See supra* note 7 for an analogous AID situation and judicial holding.
40. The only recorded "mind change" case covered the surrogate's desire to keep her biological child—with no issue of support being raised. *Time*, June 22, 1981 at 71; *Infra*, note 41.

   See generally Curtis, The Psychological Parent Doctrine in Custody Disputes Between Foster Parents and Biological Parents, 16 *Colum. J. L. & Soc. Probs.* 149 (1980).
41. The first known incident of a surrogate mother attempting to rescind her contract was decided by a Superior Court Judge in Los Angeles, California. The plaintiff and his wife were unable to have a family and therefore sought a California widow and mother of three children to be put under contract to be artificially inseminated by plaintiff. Although not paid for her services, her medical expenses were covered. During the pregnancy, the surrogate changed her mind and expressed her intent to keep the fetus when born. Plaintiff then sued for custody but before trial requested the presiding judge to withdraw his suit. Claiming "extraordinary publicity" of the fact that his wife was a transsexual would make it difficult for his son to "lead a normal life," the plaintiff capitulated. The infant was given his birth mother's surname, but the plaintiff was listed on the birth certificate as the father and was granted visitation rights. The plaintiff's attorney opined that plaintiff might well wish to reopen the case on the issue of visitation rights. The presiding judge of the court stated his belief that surrogates should always be allowed to reconsider and change their minds regarding such contracts. *Wash. Post*, April 7, 1981, at A7, cols. 1–2; *Wash. Post*, June 5, 1981, at A6, col. 1; *Time*, June 22, 1981, at 71; Mathews, Adoptive Parents Fight Surrogate Mother for Baby, *Wash. Post*, Mar. 23, 1981, at 7A, col. 6.
42. *Time*, June 5, 1978, at 59.

   What if a biologic mother decides to have an abortion owing to the discovery of fetal abnormality? Would any action be available to the contracting parties to prevent such a decision by the surrogate? Paton v. Trustees of BPAS [1978] 2 All E. R. 987 has, by implication, answered to this conundrum. There it

was held that a man could not prevent his wife from having
an abortion. It could thus be argued that the biological mother
has as much standing vis-à-vis the surrogate mother as the father vis-à-vis the mother when considering undergoing an
abortion. Brown, Legal Implications of The New Reproductive
Technology in *Making Babies: The Test Tube and Christian Ethics*
61, 67 (A. Nichols & T. Hogan eds. 1984).

43. Doe et al v. Kelley, State Attorney General, Jan. 1980 *Reporter
on Human Reproduction and the Law*, II-B-15–II-B-22.
44. Mich. Comp. Laws Ann., §§ 710.54, 710.69 (1980 Supp.).
45. *Supra* note 43, at 16–17.
46. *Id.* at 18.
47. *Id.* at 19, 20.
48. *Id.* at 19.
49. *Id.* at 21.
50. *Wash. Post*, April 24, 1981 at A12, col. 1.

The very first recorded incident of a surrogate motherhood
birth took place in San Francisco, California, on September 6,
1976. N. Keane & D. Breo, *The Surrogate Mother* 33 (1981).
51. *Id.*
52. Ky. Rev. Stats. § 199.590 (1977 Rpl.).

In addition to his advisory opinion, 80-18, the Attorney
General maintained an action to enjoin Surrogate Planning As-
sociates, Inc. from making any further surrogate mother ar-
rangements in the state. Annas, Contracts to Bear a Child;
Compassion or Commercialism, *Hastings Center Rep.* 23, 25
(April 1981).

Ms. Kane received less than $10,000 for her services. The
issue born of this arrangement thus became the first born
through the work of Surrogate Parenting Associates (SPA)—
an organization founded in January, 1980, in Louisville, Kentucky, whose function is to be a matchmaker—matching, as
such, infertile couples with fertile women willing to bear babies. SPA estimates the average cost involved in a total surrogate program to be anywhere from $13,000 to $20,000, which
covers medical and hospital expenses and the fees of SPA. Krucoff, Private Lives: The New Surrogate, *Wash. Post*, Sept. 24,
1980 at B5, col. 2. Presumably, a monetary gift or other per-
sonalty or real property could be made to a surrogate, however,
for her undertaking and not be voided.
53. Excepts from Decision by New Jersey Supreme Court in the

Baby M Case, *N.Y. Times*, Feb. 4, 1988, at B6, col. 1. In re Baby M, 14 *Fam. L. Rep.* (BNA) 2007 (1988).

See generally Johnson, The Baby M Decision: Specific Performance of a Contract for Specially Manufactured Goods, *S. Ill. U. L. Rev.* 1339 (1987).

54. In the Matter of Baby M, No. FM-25314-86E (Superior Ct. of N.J., Chancery Div., Family Part, Mar. 31, 1987). 107 N.J. 49, 526A. 2d 138.

55. *Supra* note 53.

56. *Id.*

See Note, Developing a Concept of the Modern Family: A Proposed Uniform Surrogate Parenthood Act, 73 *Geo. L.J.* 1288 (1985).

See also Symposium on Surrogate Motherhood, 16 *Law, Med. & Health Care* 1 (1988).

57. Royster Guano Co. v. Virginia, 253 U.S. 412, 415 (1920).

58. *Id.* Accord, City of New Orleans v. Dukes, 427 U.S. 297, 303 (1976); Dandridge v. Williams, 397 U.S. 471, 485 (1970).

59. Zablocki v. Redhail, 434 U.S. 374, 383 (1978).

60. Eisenstadt v. Baird, 405 U.S. 438, 447, n. 7 (1972) (emphasis in original). *See* Shapiro v. Thompson, 394 U.S. 618, 638 (1969).

61. San Antonio School District v. Rodrigez, 411 U.S. 7, 33–34 (1973).

62. *See*, e.g., Personnel Administrator of Mass. v. Feeney, 442 U.S. 256, 272 (1978).

63. Hafen, The Constitutional Status of Marriage, Kinship, and Sexual Privacy—Balancing the Individual and Social Interests, 81 *Mich. L. Rev.* 463, 541 (1983).

64. *Id.*

See, e.g., Califano v. Jabst, 434 U.S. 47 (1977); Zablocki v. Redhail, *supra* note 59.

65. *Supra* note 63, at 551.

66. *See supra* note 2.

67. L. Tribe, *American Constitutional Law* 892 (1978).

68. *Id.* Ch. 15 at 889.

69. *Id.* at 892.

70. H. Hart, *Law, Liberty and Morality* 57 (1963).

71. *Id.*

72. P. Devlin, *The Enforcement of Morals* 25 (1965).

See Dworkin, Lord Devlin and The Enforcement of Morals, 75 *Yale L. J.* 986 (1966).

73. Shockingly, using this former view of openness in matters of private conduct, it was recently held in New York that unmarried, consenting adults in private settings should not be subjected to criminal sanctions for acts of sodomy since these actions were of no societal concern and promoted no social damage. People v. Onofre, 51 N.Y.2d 476, 415 N.E.2d 936, 434 N.Y.S.2d 947 (1980); cert. denied, 451 U.S. 987 (1981).

74. 274 U.S. 200 (1927).

75. *Id.* at 207.

76. Comment, *supra* note 21, at 1054.

77. *Id.*

78. Note, Legislative Naivete in Involuntary Sterilization Laws, 12 *Wake Forest L. Rev.* 1064, 1071 (1976). Writing for the Court, Justice Holmes observed:

> We have seen more than once that the public welfare may call upon the best citizens for their lives. It would be strange if it could not call on those who already sap the strength of the State for these lesser sacrifices, often not felt to be such by those concerned, in order to prevent our being swamped with incompetence. It is better for all the world, if instead of waiting to execute degenerate offspring for crime, or to let them starve for their imbecility, society can prevent those who are manifestly unfit from continuing their kind.

Buck v. Bell, *supra* note 74, 274 U.S. at 207.

79. 316 U.S. 535 (1942).

80. The Bell Court has used a revolving door rationale in rejecting the equal protection claim:

> [T]he law does all that is needed when it does all that it can, indicates a policy, applies it to all within the lines all similarly situated so far and so fast as its means allow. Of course so far as the operations enable those who otherwise must be kept confined to be returned to the world, and thus open the asylum to others, the equality aimed at will be more nearly reached.

Buck v. Bell, *supra* note 74, 274 U.S. at 208.

81. Skinner v. Oklahoma, *supra* note 79, 316 U.S. at 537.

82. *Id.* at 541.

83. *See,* e.g., Cleveland Bd. of Ed. v. La Fleur, 414 U.S. 632, 640 (1974).

84. Comment, *supra* note 21, at 1056. Indeed, one commentator has suggested that this case has been incorrectly interpreted since "the Skinner Court neither denied the state's right to ster-

ilize nor established a constitutional right to procreate. Rather, the Court expressly declared that the scope of the state's police power was unaffected by its holding." *Id.*

85. *See* Relf v. Weinberger, 372 F. Supp. 1195, 1999 (D.D.C. 1974).
86. Skinner v. Oklahoma, *supra* note 81, 316 U.S. at 536. Writing for the court, Justice Douglas stated:

> The power to sterilize, if exercised, may have subtle, far-reaching and devastating effects. In evil or reckless hands it can cause races or types which are inimical to the dominant group to wither and disappear. There is no redemption for the individual whom the law touches.

Id. at 541.

87. *Supra* note 63, at 538.
88. 381 U.S. 479 (1965).
89. The Court observed that "specific guarantees in the Bill of Rights have penumbras, formed by emanations from the guarantees that help give them life and substance." *Id.* at 484 (citation omitted). Thus, it was those "[v]arious guarantees [which] created the zones of privacy." *Id.*
90. Eugenic Artificial Insemination: A Cure for Mediocrity? 94 *Harv. L. Rev.* 1850, 1853 (1980) [hereinafter Developments in the Law].
91. *Id.*
92. Comment, *supra* note 21, at 1058.
93. Griswold V. Connecticut, *supra* note 88, 381 U.S. at 498–499.
94. 405 U.S. 438 (1972).
95. *Id.* at 453 (emphasis in original).
96. *Id.* at 485.
97. "It has been suggested that the Court's opinion was lacking in candor, for it stated in broad dictum a major extension of the 'privacy right' which could have justified its decision, while purporting to rest on a strained conclusion that the statute involved failed even the minimal rationality test." Developments in the Law, *supra* note 90, at 1184 (footnotes omitted).

Under an expansive liberal interpretation, Eisenstadt has been held to extend the right of privacy to all sexual activities of whatever nature. *See* Wilkinson & White, Constitutional Protection for Personal Life Styles, 62 *Cornell L. Rev.* 563 at 589 (1977). *See*, e.g., Lovisi v. Slayton, 363 F. Supp. 620, 625 (E.D. Va. 1973), aff'd, 539 F.2d 349 (4th Cir.) (*en banc*), cert. denied, 429 U.S. 977 (1976).

A more conservative and narrow construction views Ei-senstadt as merely recognizing a freedom to decide issues related to the birth of a child. *See*, e.g., State v. Santos, 413 A.2d 58, 68 (R.I. 1980).

98. 410 U.S. 113 (1973).

99. *Id.* at 153. This right, however, was not absolute and the degree of involvement allowed would be contingent on the length of the pregnancy. "[P]rior to approximately the end of the first trimester, the abortion decision and its effectuation must be left to the medical judgment of the pregnant woman's attending physician." *Id.* at 164. After this stage, the "State may regulate the abortion procedure to the extent that the regulation reasonably relates to the preservation and protection of maternal health." *Id.* Finally, after viability, the state may protect fetal life and "may go so far as to proscribe abortion during that period, except when it is necessary to preserve the life or health of the mother." *Id.* at 163–164.

100. In support of this proposition the Court cited Buck v. Bell, *supra* note 74, which led one commentator to observe: "As it is difficult to imagine a more substantial interference with procreation than compulsory sterilization, the limited nature of the recognized procreative 'right' is apparent." Developments in the Law, *supra* note 90, at 1868.

101. 431 U.S. 678 (1977) (plurality opinion).

102. In addition to the privacy cases already discussed, the Court cited Cleveland Board of Education v. LaFleur, 414 U.S. 632 (1974); Loving v. Virginia, 388 U.S. 1 (1967); Prince v. Massachusetts, 321 U.S. 158 (1944); Pierce v. Society of Sisters, 262 U.S. 390 (1923).

103. Carey v. Population Servs. Int'l, *supra* note 101, 431 U.S. at 685.

104. Kritchevsky, *supra* note 2, at 27–28.

105. Carey v. Population Servs. Int'l, *supra* note 101, 431 U.S. at 688 (emphasis added).

106. *Id.* at 688, n. 5. *See generally* Paris Adult Theatre I v. Slaton, 413 U.S. 49, 68, n. 15 (1973) (implication that state fornication statutes do not violate the federal constitution). *But see* State v. Saunders, 75 N.J. 200, 381 A.2d 333 (1977) (holding that fornication statute involves by its very nature a personal choice and that it infringes upon the right of privacy).

107. Zablocki v. Redhail, 434 U.S. 374, 387 (1977); cf. Doe v. Commonwealth Attorney, 403 F. Supp. 1199 (E.D. Va. 1975), aff'd,

425 U.S. 901 (1976) (summary affirmance of three-judge district court decision holding that the state of Virginia could constitutionally apply its sodomy statute to private sexual conduct between consenting male adults). *See* Bowers v. Hardwick, 478 U.S. 186, 106 S. Ct. 2841 (1986) where a 5-4 U.S. Supreme Court upheld a Georgia statute that makes it a crime to engage in private, consensual sodomy.

108. Developments in the Law, *supra* note 91, at 1185. *See also* Wilkinson & White, Constitutional Protection for Personal Life Styles, 62 *Cornell L. Rev.* 563, 591–594 (1977).

109. Maher v. Roe, 432 U.S. 464, 478 (1976).

110. Lindsey v. Normet, 405 U.S. 56, 74 (1972).

111. Massachusetts Bd. of Retirement v. Murgia, 427 U.S. 307, 314 (1975).

112. Weber v. Aetna Cas. & Sur. Co., 406 U.S. 164, 173 (1972).

113. Maynard v. Hill, 125 U.S. 190, 211 (1888).

114. Zablocki v. Redhail, 434 U.S. 374, 386 (1977).

115. *See generally* Parham v. J.R., 442 U.S. 584 (1979); Wisconsin v. Yoder, 405 U.S. 205 (1972); Pierce v. Society of Sisters, 268 U.S. 510 (1925); Meyer v. Nebraska, 262 U.S. 390 (1923).

116. *See, e.g.,* Ferguson v. Finch, 310 F. Supp. 1251 (D.S.C. 1970).

117. Kritchevsky, *supra* note 2, at 31.

118. 69 Misc. 2d 304, 330 N.Y.S.2d 235 (1972).

119. Id. at 314, 330 N.Y.S.2d at 245; cf. Smith v. Organization of Foster Families for Equality & Reform, 431 U.S. 816 (1977).

120. Kritchevsky, *supra* note 2, at 29.

121. Smith v. Organization of Foster Families for Equality & Reform, *supra* note 119, 431 U.S. at 843.

122. Developments in the Law, *supra* note 90, at 1270.

123. Kindregan, State Power Over Human Fertility, 23 *Hastings L. J.* 1401, 1409 (1972).

124. Griswold v. Connecticut, *supra* note 93, 381 U.S. at 486.

125. Karst, The Freedom of Intimate Association, 89 *Yale L. J.* 624 (1980).

126. 431 U.S. 494 (1977).

127. Developments in the Law, *supra* note 90, at 1272.

128. *Id.* at 1271.

129. Dandridge v. Williams, 397 U.S. 471, 485 (1969) quoting Lindsey v. Natural Carbonic Gas Co., 220 U.S. 61, 78 (1910).

   But see M. Glendon, *The New Family and the New Property* (1981).

130. Hafen, The Constitutional Status of Marriage, Kinship, and Sexual Privacy—Balancing the Individual and Social Interests, 81 *Mich. L. Rev.* 463, 559 (1983).
131. May v. Anderson, 345 U.S. 528, 536 (1952) (Frankfurter, J., concurring).
132. *Supra* note 52.
133. R. Scott, *The Body as Property*, 221 (1981).
134. Brown, Legal Implications of The New Reproductive Technology in *Making Babies: The Test Tube and Christian Ethics*, at 61 (A. Nichols & T. Hogan eds. 1984).
135. *Id.*
136. *Supra* note 134, Foreword at vii, ix.
137. *Id.*
138. *See* Brumby, Australian Community Attitudes in *In-Vitro* Fertilization, *Med. J. Australia* 650 (Dec. 10–24, 1983).
139. Asche, *supra* note 26.
140. 138 Mass. 14 (1884).
141. 33 Cal. App. 2d 629, 92 P. 2d 678, aff'd *per curiam*, 33 Cal. App. 2d 629, 93 P. 2d 562 (1939).
142. 65 F. Supp. 138 (D.C. Cir. 1946).
143. 282 A.D. 542, 125 N.Y.S. 2d 696 (1953).
144. *Id.*
145. *See* Roller v. Roller, 37 Wash. 242, 79 P. 788 (1905).
146. Comment, Surrogate Mothers and Tort Liability: Will the New Reproductive Technologies Give Birth to a New Breed of Prenatal Tort?, 34 *Clev. St. L. Rev.* 311, 322–323 (1986).

    See, e.g., Attwood v. Estate of Attwood, 276 Ark. 230, 633 S. W. 2d 366 (1982).

    See also Robertson, Procreative Liberty and the Control of Conception, Pregnancy and Childbirth, 69 *Va. L. Rev.* 405, 447–450 (1983); Comment, The Reasonable Parent Standard: An Alternative to Parent-Child Tort Immunity, 47 *U. Colo. L. Rev.* 795 (1976).
147. People v. Yates, 114 Cal. App. Supp. 782, 298 P. 961 (1931) (*per curriam*). Accord, People v. Sianes, 134 Cal. App. 355, 25 P. 2d 487 (1933).
148. *See generally* Comment, Redefining Mother: A Legal Matrix for New Reproductive Technologies, 96 *Yale L. J.* 187 (1986).
149. *See* Day v. Nationwide Mutual Ins. Co., 328 S. 2d 560 (Fla. Dist. Ct. App. 1976).
150. Restatement 2d, Torts § 430 comment d (1965).

151. Seavey, Negligence—Subjective or Objective? 41 *Harv. L. Rev.* 1, 18 (1927).
152. Comment, Surrogate Mothers and Tort Liability: Will the Reproductive Technologies give Birth to a New Breed of Prenatal Tort?, 34 *Clev. State L. Rev.* 311 at 346 (1986); Note, Maternal Substance Abuse: The Need to Provide Legal Protection for the Fetus, 60 *S. Cal. L. Rev.* 1209 (1987); Case Study and Commentary, When a Pregnant Woman Endangers Her Fetus, *Hastings Center Rep.* 24 (Feb. 1986).

   See also Shaw, Should Child Abuse Laws be Extended to Include Fetal Abuse? in *Genetics and the Law* III at 309 (A. Milunsky & G. Annas eds. 1985).

CHAPTER 10

1. D. Hancock & D. Ford, Frozen Embryo Orphans Heir to $8m Estate, *The Australian*, June 18, 1984, at 7, col. 7.

   In 1983, Canberra, Australia, became the first place in the world where a human pregnancy resulted from a fertilized egg that had been frozen, thawed, and implanted in the mother's womb. Branigan, Frozen Embryos Trigger Debate by Australians, *Wash. Post*, May 17, 1983, at 1, col. 6. The costs for developing and maintaining an in vitro or test-tube baby program for 7 years in order to produce Australia's first progeny therefrom cost over $1 million. Williamson, Test Tube Births: An Uncertain Way Ahead, *The Nat'l Times*, Feb. 7–13, 1982, at 12, col. 5.
2. C. Wallis, Quickening Debate Over Life on Ice, *Time*, July 2, 1984, at 46.
3. Lawson, Molloy, Jobson, & Walley, The Frozen Embryo Mystery: The Life and Strange Times of Elsa Rios, *The [Australian] Bulletin*, July 3, 1984, at 22.
4. Gavigan, The Criminal Sanction as It Relates to Human Reproduction: The Genesis of The Statutory Prohibition of Abortion, 5 *J. Legal History* 20 (1984).
5. *Id.*
6. *Id.* at 21.
7. *Id.*
8. *Ethics in Medical Research Involving the Human Fetal Tissue*, para's. 2.9–2.10 (Canberra, 1983).

9. Bates, Legal Criteria for Distinguishing Between Life and Dead Human Fetuses and Newborn Children, 8 *U. New S. Wales L. Rev.* 143 (1983).

10. McKay v. Essex Health Auth., 2 W.L.R. 890 (1982).
    See Reagan, Abortion and The Conscience of The Nation, 9 *Human Life Rev.* 7 (1983).

11. Atty. Gen. for The State of Queensland (Ex rel. Kerr) and Arthur v. T., 57 *Australian L.J. Rev.* 385 (1983). This would be, essentially, the same posture of the United States courts as a consequence of Roe v. Wade, 432 U.S. 464 (1977).
    But see M. Tooley, *Abortion and Infanticide* (1983).

12. Lucas, Abortion in New South Wales—Legal or Illegal, 52 *Australian L. J.* 327–339 (1978).

    For a comparative analysis of the American and British posture here *see* Annas & Elias, *In Vitro* Fertilization and Embryo Transfer: Medico-Legal Aspects of a New Technique to Create a Family, 17 *Fam. L. Q.* 199 (1983); Edwards & Steptoe, Current Status of IVF and Implantation of Human Embryos, *The Lancet* 1265 (Dec. 1983).

13. *Supra* note 9, at 42–43.

14. *Id.*

15. *Supra* note 1.

16. *Id.*

17. *Supra* note 1.

    It was determined by the Victorian government on December 3, 1987, that the frozen embryos will be thawed and, if still "alive," given to a childless couple. Should subsequent birth and survival occur, no share in the inheritance of the Rios' estate will pass to the offspring. Dalton, Dead Couple's Embryos to be Thawed, *Wash. Post*, Dec. 4, 1987, at A38, col. 1.

18. *See* W. Raushenbush, *The Law of Personal Property* §§ 1.5, 1.7 (3rd ed. 1975).

19. *Id.*

20. G. C. Bogert & G. T. Bogert, *Handbook of the Law of Trusts* 287, 305 (5th ed. 1973).

21. M. O'Neill & B. Hutton, Advice Exists on Rios Embryos, Says Health Council, *The [Australian] Age*, July 4, 1984, at 10, col. 3.

22. *Sydney Sunday Telegraph*, Nov. 7, 1982, at 6, col. 1.

    The South Australian government gave its approval in April, 1984, to the freezing of fertilized ova—all as part of its test tube baby procedures. It specifically prohibited the use of frozen fertilized ova for scientific or genetic research and the practice of surrogate. [*Sydney*] *Sunday Morning Herald*, April 23, 1984, at 9, col. 5.

23. R. Scott, *The Body as Property* 202 (1982).

See Proceedings of the Conference, In Vitro Fertilization: Problems and Possibilities, March 11, 1982, Monash University, Center for Human Bioethics, Melbourne, Australia.

24. M. Kirby, Foreword, *Making Babies: The Test Tube and Christian Ethics at ix* (1984).

25. *Supra* note 3, at 25.

26. Sec. 2.16.

27. Sec. 2.19.

28. Sec. 2.18.

See Orphaned Embryos May Be Left to Thaw, *Sydney Morning Herald*, Sept. 4, 1984, at 3, col. 1.

29. Sec.'s., 3.25–3.28.

30. Sec. 3.29.

31. See Victoria Will Bar Payments to Surrogate Mothers, *Sydney Morning Herald*, Sept. 4, 1984, at 3, col. 2.

What could be regarded as the British counterpart of the Waller Committee—the Warnock Commission—proposed a ban on surrogate mothers, yet concluded that embryo research should be permitted until the fourteenth day after fertilization when the first identifiable features of the embryo develop. *Time*, Aug. 6, 1984, at 50.

See Report of the Committee of Inquiry into Human Fertilization and Embryology (July 1984) The Warnock Report).

See generally Smith, Intimations of Life: Extracorporeality and the Law, 21 *Gonz. L. Rev.* 395 (1986).

32. *Supra* note 3, at 25.

33. Australians Reject Bid to Destroy 2 Embryos, *N.Y. Times*, Oct. 24, 1984, at A 18, col. 1.

34. In an interview conducted by David Hartman of the ABC Television Network, *Good Morning America's* program of October 24, 1984, with Mr. Hayden Storey, the author of the legislation in Victoria, Mr. Storey stressed that his proposal had only specific application to the "special category" which the Rios' embryos enjoyed and that he anticipated no extension to other possible cases.

35. Corns, Legal Regulation of In Vitro Fertilization in Victoria, [*Victoria*] *Law Institute Journal* 838 (July 1984).

36. *Supra* note 3, at 25.
37. Asche, Ethical Implication on the Use of Donor Sperm, Eggs and Embryos in The Treatment of Infertility, 57 [*Australian*] *Law Institute J.* 716–719 (1983).
38. Nossal, The Impact of Genetic Engineering on Modern Medicine, 27 *Quandrant* 22–27 (Nov. 1983).
39. Annas, Redefining Parenthood and Protecting Embryos: Why We Need New Laws, *Hastings Center Rep.* 50 (Oct. 1984).
40. For an eloquent analysis of legislative, as opposed to judicial strengths in managing the problems of in vitro fertilization, *see—* from Sydney, Australia—Justice Michael D. Kirby's address, IVF—The Scope and Limitation of Law," presented in London, England, September, 1983, at The Conference of Bioethics and The Law of Human Conception—*In Vitro* Fertilization.
41. *See Test-Tube Babies: A Guide to Moral Questions, Present Techniques* (W. Walter & P. Singer eds. 1982).

CHAPTER 11

1. Hoagland, Some Reflections on Science and Religion, in *Science Ponders Religion* 17, 18 (H. Shapley ed. 1960) (quoting the physicist P.W. Bridgman).
2. B. Russell, *The Scientific Outlook* 273 (1931).
3. B. Russell, *The Impact of Science on Society* 98 (1952).
4. *Id.* at 29. The Greeks, with Archimedes being the exception, were interested only in the first function. The Arabs, however, were in quest of the elixir of life and the methods needed to transmute base metals into gold. *Id.*
5. *Id.* at 98. During the past three centuries, the science which has been rated as successful has consisted "in a progressive mathematization of the sensible order . . ." *Id.*
   The history of science reveals that it is based on creative leaps of imaginative vision. L. Gilkey, *Religion and the Scientific Future* 45 (1970). *See* J. Maritain, *Science and Wisdom* (1940); H. Muller, *Science and Criticism* (1943).
6. Hoagland, Some Reflections on Science and Religion, in *Science Ponders Religion* 17, 24 (H. Shapley ed. 1960). The examples used for support of this last statement are: the certainty that the earth is round, not flat, and the realization that biological evolution, by natural selection, is no longer just a theory but is a high probability. *Id.*

In its fundamental phase, science is explanation by description using methods of observation and experiment. The fundamental assumptions which it makes are practical conclusions of common sense: namely, that the objects and the events constituting the material universe are in a necessary connection with one another and that man, by his decisions, can affect the order and events of the universe itself. W. Schroeder, *Science, Philosophy and Religion* 44, 45, 58 (1933).

7. J. Huxley, *Science, Religion and Human Nature* 20, 21 (1930).

8. *Id.* at 58.

9. B. Russell, *Religion and Science* 3 (1935). *See* A. Barbour, *Myths, Models and Paradigms—The Nature of Scientific and Religious Language* (1974).

10. B. Russell, *Religion and Science* 11 (135). Russell lists the fact that the historical religions have had a Church and a code of personal morals as a reason for further conflict. *Id.* at 4.

11. Russell, *supra* note 10, at 14.

    See generally Gustafson, Theology Confronts Technology and the Life Sciences, *Commonweal* 386 (June 16, 1978).

12. B. Russell, *The Impact of Science on Society* 16 (1952).

13. While religion seeks to explain the obvious in terms of mystery, science masters the simple and obvious and then witnesses, by the application of elemental principles, the dissolution of the complex. F. Northrup, *Science and First Principles* (1931).

    See also A. Whitehead, *Science and the Modern World*, ch. 13 (1926).

14. A. Whitehead, *The Interpretation of Science* 179 (A. Johnson ed. 1961) [hereinafter referred to as Whitehead]. *See also* L. Gilkey, *Religion and the Scientific Future*, at Ch. 1 (1970).

    See generally Dobzhansky, Evolution: Implications for Religion, in *Changing Man: The Threat and the Promise* 142 (K. Haselden & P. Hefner eds. 1968).

15. Evans, Rationalization, Superstition and Science, in *Science, Reason and Religion* 43, 45 (C. Macey ed. 1974).

16. *Supra* note 6, at 17.

17. Gilkey, *supra* note 14, at 4.

18. Whitehead, *supra* note 14.

19. *Id.* at 176. *See also* Gilkey, *supra* note 14.

20. C. Coulson, *Science, Technology and the Christian* 48 (1960).

21. Gilkey, *supra* note 14, at 25.

22. Gilkey, *supra* note 14, at 25.

23. Schroeder, *supra* note 6, at 61.

24. *Id.* at 62, 63.
25. Barbour, The Methods of Science and Religion, in *Science Ponders Religion* 214, 215 (H. Shapley ed. 1960).
26. Murray, Two Versions of Man in *Science Ponders Religion* 147, 48 (H. Shapley ed. 1960).
27. Burhoe, Salvation in The Twentieth Century, in *Science Ponders Religion* 65, 77, 78 (H. Shapley ed. 1960).
28. *Supra* note 23, at 60.
29. C. Miller, *A Scientist's Approach to Religion* 29, (1947).
30. Hasset, Freedom and Order Before God: A Catholic View, 31 *N.Y.U. L. Rev.* 1170, 1180 (1956).
31. H. Smith, *Ethics and the New Medicine* 64 (1970).

According to St. Augustine, a sexual act deprived of its procreative character was illegitimate. Thus, if, in the name of love, a couple chooses to express themselves sexually, they should accordingly perform the authentic sexual act not deprived of its procreative character. Love and procreation are inseparable. Smith, Theological Reflections and the New Biology, 48 *Ind. L. J.* 605, 619, 621 (1973).

See also St. John-Stevas, A Roman Catholic View of Population Control, 25 *Law & Contemp. Probs.* 445, 446 *passim* (1960).
32. *Supra* note 30, at 1179.
33. *Id.* at 1180.

Today, modern theologians and pastoral counselors would not view AIH as immoral or repulsive to marriage. Donor insemination (AID) is still regarded as violative of Catholic Church dogma. *Human Sexuality—New Direction in American Catholic Thought* 137–139 (1977).
34. *See Human Sexuality, supra* note 33, at 138–139.
35. *See* Theological Reflections and the New Biology, *supra* note 31, at 620.
36. J. Fletcher, *The Ethics of Genetic Control* 114 (1974).
37. *Supra* note 35, 621.

See generally Making Babies: The Test Tube and Christian Ethics (A. Nichols & T. Hogan eds. 1984).
38. *Supra* note 35, at 622. Use of a woman's womb by another couple would be considered by the Church as "analogous to allowing use of one's body solely for the sexual pleasure of another, and, thus immoral." *Id.* at 621.
39. *See Human Sexuality, supra* note 33, at 138–139.
40. *Id.* at 137.

41. *Id.* at 137–138.
42. *Id.* at 138.
43. *Id.*
44. *Id.* at 139.
45. A Swift Stunning Choice, *Time*, Sept. 4, 1978, at 65, 66.
46. *Id.*
47. 38 Ecumenical Courier 1, 5 (Nos. 3–4, 1979).
48. Wojtyla, *Love and Responsibility* (1960), commented on in *Time*, Oct. 30, 1978, at 94.
49. *Id.* at 97.
50. Pope Warns Against Misuse of New Medical Procedures, *Wash. Post*, Oct. 28, 1980, at A4, col. 4.

   See generally D. Kelly, *The Emergence of Roman Catholic Medical Ethics in North America* (1979).
51. H. Smith, *Ethics and the New Medicine* 66, 67 (1970).
52. *Id.* at 67.
53. *Id.*
54. *Id.* at 68.
55. *Id.* at 69, 70 *passim. See generally* In Vitro Fertilization; Four Commentaries, 8 *Hastings Center Rep.* 7 (1978).
56. Ramsey, Freedom and Responsibility in Medical and Sex Ethics: A Protestant View, 31 *N.Y.U.L. Rev.* 1189, 1198 (1956).
57. For an argument regarding the compatibility of AID with the Christian understanding of secularity, marriage and parenthood, see J. Fletcher, *Morals & Medicine* 118 (1960).
58. Rackman, Morality in Medico-Legal Problems: A Jewish View, 31 *N.Y.U.L. Rev.* 1205, 1208 (1956). Although Jewish ethics would favor experiments and tests to discern possible genetic malfunctions which would result in congenital disease before the birth of a fetus, the artificiality of test tube babies and of cloning, for example, would be disregarded as tampering too much with the basic structures of creations. Siegel, The Ethical Dilemma of Modern Medicine: A Jewish Approach, 3 *The Kennedy Inst. Q. Rep.* 5 at 7 (No. 1, 1976–1977).
59. Rackman, *supra* at 1210.
60. *Id.*

   In 1958, the Chief Rabbi of Israel, Rabbi Nissim, ruled that children born to parents as the result of artificial insemination will be recognized by the Jewish religion as legitimate. A. Scheinfeld, *Your Heredity and Environment* 665 (1965).
61. Rackman, *supra* note 58, at 1209.

62. *Id.* at 1209–1210.
63. J. Fletcher, *The Ethics of Genetic Control* 114, 115 (1974).
    See generally D. Gosling, *Science and Religion in India* (1976).
    For an interesting perspective on atheist realism and Marxist
    dialectics regarding the New Biology, *See* P. Chauchard, *Science
    and Religion*, Ch. 3 (1962). The beliefs and reactions of other re-
    ligions here may be found in: 3 *Encyclopedia of Bioethics* 901–1020,
    1365–1378 (W. Reich ed. 1978).
64. Fletcher, *supra* at 127.
65. See generally, A. Toynbee, *An Historian's Approach to Religion*
    (1956).
        Interestingly, a 1969 Harris opinion survey of some 1600
    adults throughout America relative to advances and applications
    of the New Biology, revealed a most interesting attitudinal pro-
    file. Nineteen percent of all interviewed approved of AID, while
    56% disapproved of the process. Where the only method for a
    married couple to conceive a family involved use of heterologous
    insemination (AID), 35% of those interviewed approved of the
    technique. Forty-nine percent of the men interviewed in the sur-
    vey agreed in principle with homologous insemination (AIH),
    while 62% of the women expressed their approval of allowing
    their husband's semen to be used, through artificial means of
    injection, in order to inseminate them. Smith, For Unto Us a
    Child is Born—Legally!, 45 *A.B.A.J.* 143 (1970).
66. *See* Editorials, New Vatican Instruction on Human Life and Pro-
    creation, *America*, Mar. 28, 1987 at 245.
67. *Id.* McCormick Editorial at, 247.
68. *Id.*
69. *Id.* at 248.
70. *Id.*
71. *Id.*
72. *Id.* Cahill Editorial 246, at 247.
73. *Supra* note 67. On June 27, 1988, the Michigan Legislature passed
    the Surrogate Parenting Act, which establishes surrogate par-
    entage contracts as contrary to the public policy and void. Those
    who enter into, induce, arrange, procure, or otherwise assist in
    such contracts will be adjudged guilty of a felony and fined up
    to $50,000, imprisoned up to five years, or both. *Mich. Compiled
    Laws Ann.*, § 722.851–722.863 (1988).

Index